OTHER BOOKS BY ARNO KARLEN

Napoleon's Glands and Other Ventures in Biohistory

Sexuality and Homosexuality

White Apples (fiction)

man
and
microbes

Disease and Plagues
in History
and Modern Times

Arno Karlen

A TOUCHSTONE BOOK
Published by Simon & Schuster
New York London Toronto
Sydney Tokyo Singapore

TOUCHSTONE
Rockefeller Center
1230 Avenue of the Americas
New York, NY 10020

Copyright © 1995 by Quantum Research Associates, Inc.

First Touchstone Edition 1996
Published by arrangement with Jeremy P. Tarcher, Inc. of G. P. Putnam's Sons

TOUCHSTONE and colophon are registered trademarks
of Simon & Schuster Inc.

Designed by Mauna Eichner

Manufactured in the United States of America

10   9   8   7   6   5   4

Library of Congress Cataloging-in-Publication Data
Karlen, Arno.
   Man and microbes : disease and plagues in history and modern times
/ Arno Karlen.
      p.   cm.
   Originally published: New York : G. P. Putnam's, 1995.
   Includes bibliographical references and index.
   1. Epidemics—History.   I. Title.
RA649.K37   1996
614.4'09—dc20   96-1705   CIP
ISBN 0-684-82270-9

contents

One   an epidemic of epidemics   1

Two   the first shocks   13

Three   revolutions   29

Four   splendor and plague   47

Five   ruthless cure   65

Six   the flying corpses of kaffa   79

Seven   the deadliest weapon   93

Eight   microbes reply   111

Nine   victory, it seems   129

Ten   a garden of germs   149

Eleven   an old thread, new twists   175

Twelve   inviting infection   195

Thirteen   from this time on   215

bibliography   231

index   250

an epidemic
of epidemics

A prediction; sadly, it comes true.
Illusions of a stainless future.
New ills and vicious changelings.
Where do new diseases come from?
We survived them before.

This book is about new plagues, survival, and the dance of mutual adaptation we carry on with our microbial fellow travelers.

An alarming tide of new and resurgent diseases has been rising around the world for decades. Now it advances faster than ever. This signals a crisis in the history of the human species. We have brought it on by rending the fabric of our environment, changing our behavior, and ironically, by our inventiveness in increasing the length and quality of our lives.

Until quite recently, few people seemed aware of this epidemic of epidemics. Even among doctors and researchers, concern was rare.

Almost twenty years ago, I told friends that I was thinking of writing about why so many new diseases were appearing. Most of my friends were puzzled. A few asked if I meant Legionnaires' disease and Lyme disease, both of which had lately appeared. I said yes, those and many others. I began gathering material.

Five years later, I wanted to write a book about the emergence of slow viruses, and the major epidemics I felt sure would follow. AIDS had not yet been identified and named. No publisher was interested. I was told that this could interest only specialists.

In 1990, I outlined this book. Its central idea still puzzled many people. When I said it was about new diseases and where they come from, most said, "You mean AIDS." A few, with more than average interest in such things, mentioned Lassa, Ebola, and Marburg fevers. Yes, I said, and many, many more.

When I began work on the book, I predicted to its publisher that before it appeared, at least two or three new epidemic diseases would make headlines; so would several old ones once thought defeated but now out of control again. Since then, a lethal hantavirus has appeared in the Four Corners region of the American Southwest. A new strain of *Escherichia coli* bacteria has caused widespread illness and death, raising worries about the safety of the nation's food supply. There have been upsurges in cases of measles, drug-resistant tuberculosis, diphtheria, and cholera. The appearance of a new, highly virulent strain of cholera bacilli may have marked the start of another global epidemic. And pneumonic plague broke out for the first time in a hundred years, in India.

For each new disease known to the general public, there are a dozen others; the wheels of biological change keep turning faster. The shared evolution of humans and microbes has accelerated to a frenzied pace, because of changes we have made in our environment and our lifestyles. Much has been written about AIDS, but far less about other new diseases. The scientific and historical research is fragmented, like pieces of a mosaic rarely assembled in more than bits and patches. In the past few years, a handful of books for nonspecialists have appeared, about one disease or another, about emerging viruses and the increase of microbial resistance to drugs. But without seeing the larger evolutionary picture, we cannot respond wisely to these challenges to our health and survival. The cost in suffering and deaths will be devastating.

We have been slow to understand that we live in a new biocultural era. For decades we cherished the myth that infectious diseases were fading forever. This was a posture born of inherited optimism. The nineteenth century generated an almost religious faith in social, scientific, and technological progress. Such optimism enabled people to call the slaughter of 1914–1918 a war to end all wars. The two great global epidemics of that era, typhus and type A influenza, each killed 20 million people or more, dwarfing the toll of combat, without blunting popular faith in medical progress.

At first, events seemed to justify such optimism. More than a half-century would pass before the arrival of AIDS, another epidemic that kills by the millions. Those decades brought cleaner food and water, better living conditions, polio vaccine, antibiotics, the eradication of smallpox, and huge reductions of such killers as tuberculosis, cholera, and syphilis. Agricultural abundance soared in the developed world, and then in developing nations. The deaths of infants and small children, for so long routine events, became unusual tragedies. Life spans increased, and in some places almost doubled. World War II was the first major war in which epidemics took far fewer casualties than battle.

Now, just a few generations later, it is difficult to appreciate how astonishing all this was. For 10,000 years, since the first hunter-gatherers settled in villages, infections had killed more people than war and famine. Suddenly, by the 1930s, a new era of fitness and longevity was arriving; to many, it promised an end to all infectious disease. Pestilence, the Fourth Horseman of the Apocalypse, which a generation earlier had galloped the world, seemed a quaint bogey. Cities of the future were portrayed sparkling under plastic bubbles no germ or poison could penetrate. From now on, medicine's great task would be treating cancer, clogged arteries, stress, and the other so-called diseases of civilization and of aging. When fluoridation arrived, after World War II, it seemed that we would even reach vigorous old age with our teeth intact.

In 1969, the U.S. surgeon general, Dr. William H. Stewart, told the nation that it had already seen most of the frontiers in the field of contagious disease. Epidemiology seemed destined to become a scientific backwater. A decade later, the governments of the United States, Canada, and Great Britain announced that their citizens must recognize a radical transformation; they were threatened no longer by microbes but by their own heedlessness. Drinking, smoking, and driving without seat belts had replaced bubonic plague, smallpox, and cholera. The authorities had a point, but they failed to mention such expanding avenues of infection as drugs, sex, rapid world travel, new medical procedures, and environmental degradation.

In fact, as developed nations indulged visions of an antiseptic age, new diseases were already appearing. At first they arose mostly in remote parts of Africa and Asia. Some took small tolls; other caught the eyes of only the dwindling ranks of specialists in tropical medicine and epidemiology. In the 1960s and 1970s, new diseases struck with greater impact and visibility. Old ones reappeared as vicious changelings, resistant to drugs that had once controlled them. Syphilis, malaria, and measles made frightening comebacks. Even bubonic plague popped up sporadically in Sun Belt suburbs, and among American troops serving in Vietnam.

A few scientists voiced urgent worry, but they went largely unheeded. Most of their colleagues, like most of the public, still thought the world was advancing into a golden biomedical era. Some predicted healthy life spans in triple digits. In retrospect, such confidence seems not only foolhardy but arrogant.

The 1980s brought more reports of new diseases, drug-resistant bacteria, and thriving disease carriers, from mosquitoes to household pets. Infections once limited to small areas were spreading; Lyme disease and Rocky Mountain spotted fever, once local concerns, ranged from coast to coast. AIDS became, inevitably, a national obsession. At the same time, a trickle of environmental warnings swelled to a dire stream. Pollutants, carcinogens, and ecological insults from the groundwater to the ozone layer were said to threaten the health of the entire biosphere. Yet few people made the connection between environmental and epidemiological changes.

In the 1990s, we can see that for each disease conquered, another has emerged or reemerged. Scores of infections have shattered the dream of a sanitary utopia, where only genetic controls on aging limit the human life span.

Epidemics are again a regular part of the news. The genital herpes virus infects half the people in the United States. Chlamydia, virtually unknown until twenty years ago, has become the country's most common infectious disease after the common cold. Germs that used to attack cats, rats, sheep, and monkeys have sickened people from Albu-querque to Moscow. Many forms of cancer are more common, and viruses are implicated in helping to cause several types. Viruses are also suspected of playing roles in chronic fatigue syndrome, Alzheimer's disease, rheumatoid arthritis, systemic lupus, and multiple sclerosis. In recent years, syphilis, tuberculosis, measles, whooping cough, and diph-theria have surfaced not only in poor but in developed nations.

This has all happened with amazing speed—in evolutionary time, during the briefest blink of an eye. The common failure to ask why it has happened reflects a normal tendency toward self-protection. The story of an individual tragedy may grab our attention and threaten our comfort, but we are free to turn our backs when we feel uneasy. News of epidemics and broad biosocial change is different; it brings lingering discomfort. We fear, at the least, an attack of medical student syndrome—the dread, while reading about illness, that one's every inner twitch reflects a malady beyond cure. At the worst, the news makes us imagine cataclysm, and we want to flee the subject before anxiety or fatalism catches up to us.

Ignorance, however, is a destructive luxury when infections again threaten to take more lives than war and famine. Such a crisis could come during the next fifty years, shaping our lives and our children's. Only by understanding it can we hope to slow and change its course. To reach that understanding requires a fresh look at our relationship with the rest of the living world. And that is the aim of this book.

The table of new and resurgent diseases on page 6 is not complete, but it suggests their number and variety. As it shows, warnings appeared in the 1950s, when new hemorrhagic fevers broke out in places as far apart as Argentina and India. Hemorrhagic fevers are viral diseases that can cause internal bleeding, shock, and death; the worst are among the deadliest human infections. When the Junin virus first erupted in Argentina, in 1953, it killed up to one-fifth of its victims. In 1955, another fever, only somewhat less lethal, appeared in the Kyasanur Forest region of southwest India. Milder new viral diseases also made debuts, from O'nyong-nyong fever in Uganda, in 1959, to Oropouche in Brazil, in 1961.

These epidemics received little notice in developed nations, but it was different in 1967, when a terrifying new hemorrhagic fever leaped out of Zaire and Sudan. It seized headlines in the West when it struck

*A partial list of new diseases and the years of their first
appearance or recognition. Isolated cases or localized epidemics
of some had occurred earlier.*

Korean hemorrhagic fever (Seoul hantavirus)	1951
Dengue hemorrhagic fever	1953
Argentine hemorrhagic fever	1953
Kyasanur Forest disease	1956
Chikungunya	1955
Human babesiosis	1957
O'nyong-nyong fever	1959
Bolivian hemorrhagic fever	1960
Oropouche	1961
LaCrosse encephalitis	1965
Marburg disease	1967
Intestinal capillariasis	1967
Pontiac fever	1968
Lassa fever	1969
Human toxoplasmosis	1970
Lyme disease	1975
Ebola fever	1976
Legionnaires' disease	1976
Adult T-cell leukemia	1977
Rift Valley fever	1977
Toxic shock syndrome	1980
AIDS	1981
Eschericia coli 0157:H7	1982
Brazilian purpuric fever	1984
Human ehrlichiosis	1986
Venezuelan hemorrhagic fever	1989
Toxic-shock-like syndrome	1989
Hantavirus pulmonary syndrome	1993

*A partial list of diseases that have revived or become widespread
after previously having been localized, limited, or controlled.*

Chlamydia	Malaria
Cholera	Measles
Cervical cancer	Pertussis
Diphtheria	Pneumonic plague
Genital herpes	Syphilis
Giardiasis	Tuberculosis
Hepatitis (viral)	Viral encephalitis

thirty-one workers in a research laboratory in Marburg, Germany, and killed seven of them. Marburg disease broke out again in Africa in 1976, and its spread remains a threat in the minds of epidemiologists everywhere.

New diseases were also appearing in developed nations. In 1957, a malaria-like infection of cattle called babesiosis turned up in humans in Yugoslavia. It surfaced again in 1969 on Nantucket Island and in the late 1980s in Connecticut; its progress has not ended. In 1968, an unexplained flulike epidemic occurred in Pontiac, Michigan. Almost a decade later it would be recognized as a mild relative of another new illness, Legionnaires' disease.

Far more serious was the growing range and variety of viral encephalitis in the Americas. Many of its victims are children, and those it does not kill may be left blind, deaf, or retarded. Several types of encephalitis are transmitted from birds and wild mammals to people by mosquitoes. In the 1960s, a common new type, LaCrosse encephalitis, was identified, named for the Wisconsin city where the virus was first found. Encephalitis is sporadic, but so dangerous that health agencies monitor it rigorously. Encephalitis viruses and their carriers continue to spread to larger areas, and work and recreation patterns expose more people to the microbes. In 1990, Florida had its worst outbreak in thirty years of one type, St. Louis encephalitis. More than 200 people fell sick there and elsewhere in the United States, and nine died.

When Lassa fever, a horrid new hemorrhagic fever, appeared in Nigeria in 1969, the entire world paid attention. High-speed travel had created a global village for pathogens. Tourists brought the Lassa virus by airplane to Chicago, Toronto, and London, causing scare headlines but only isolated cases. The virus again made news in 1989. A Chicago resident flew to Nigeria for a relative's funeral; he returned and was himself buried two weeks later, killed by the Lassa virus. Like Marburg disease, Lassa fever left health departments around the world watchful and vulnerable.

When Lyme disease, a debilitating tick-borne infection, was identified in 1975, it was not, strictly speaking, new. Scattered cases of a similar ailment had been reported in Europe for many decades, but they remained little more than curiosities. Only in the mid-1970s, in Old Lyme, Connecticut, did the disease become common. Then it increased everywhere. Now it ranges from New England to California, is more frequent in Europe than in North America, and appears on other continents as well. In 1994, it was the most common of all tick-borne and insect-borne diseases in the United States.

When Ebola fever appeared, it made Lyme disease seem trivial. This hemorrhagic fever has death rates of 50 or even 90 percent. After terrifying epidemics in Africa in 1976 and 1979, it seemed to spread no further. Then in 1989 a close relative of Ebola virus turned up in monkeys imported to the United States for medical research. There were no human casualties, but the federal Centers for Disease Control (CDC) issued strict new rules on the importation, quarantine, and handling of primates, which are brought into this country by the tens of thousands each year. So far the African hemorrhagic fevers occur only sporadically outside their native territories, but some may be expanding their ranges. The frightening possibility remains that they could travel and adapt to new hosts in other regions.

In 1976, the first year of Ebola fever, the CDC was worried about a domestic problem as well, a new strain of swine flu virus that might prove to be as lethal as the one that caused the epidemic of 1918. The CDC and drug companies went on an emergency footing to produce a vaccine, and 50 million Americans received shots. The CDC was both relieved and embarrassed when flu remained relatively rare that year, even among those who received no vaccine. However, the shots caused at least 500 cases of a painful paralytic disorder called Guillain-Barré syndrome. To this day, some people avoid flu shots despite their being at high risk for such flu complications as pneumonia, because of the Guillain-Barré incident.

The swine flu scare gave Americans two warnings. One was that they still had to beware of epidemics with animal reservoirs. The flu virus has many varieties, many reservoirs (such as swine and fowl) that can exchange it, and a spectacular ability to mutate and baffle human immune defenses. The other warning was that medical technologies were creating new ills, some as threatening as those they were meant to combat.

When a severe respiratory epidemic broke out in Philadelphia in 1976, it was first feared to be swine flu. Many of the victims were American Legionnaires attending a bicentennial-year convention there; hundreds fell ill, and dozens died. Yet no flu virus was found, nor was any other familiar microbe. The mysterious ailment, dubbed Legionnaires' disease, had prominent press coverage for almost six months as researchers sought its cause. They finally found it, a peculiar bacterium that manages to thrive in air conditioners, cooling towers, whirlpool baths, and other places hostile to most life forms. Legionellosis still occurs around the world, especially in hospitals and hotels, and recently on a luxury cruise ship. There may be as many as 50,000 cases a year in the United States alone.

The appearance of new infectious diseases was no longer startling in 1980, when toxic shock syndrome (TSS) became epidemic. Hundreds of American women fell gravely ill; some died, and many suffered lasting after-effects. The cause was a toxin produced by a new, probably mutant form of a common bacterium, *Staphylococcus aureus*. TSS was linked to a new type of menstrual tampon, which subsequently was removed from the market, and few cases occurred afterward.

The next decade brought a similar disease, toxic-shock-like syndrome (TSLS). Most people first heard of it in 1990, when it killed Muppeteer Jim Henson. TSLS is extremely virulent and fast-acting; doctors said that starting antibiotic treatment just a few hours earlier might have saved Henson's life. TSLS is caused by type A streptococcus, which usually causes strep throat and scarlet fever. It declined in virulence for almost a century and now has returned in a vicious new form, probably a mutant, in the United States, Europe, and Australia. In 1994, the press made people aware that a fast-acting, supervirulent form of strep A caused "flesh-eating" infections in England and the United States, devouring patients' muscle tissue and killing some of them.

By the 1980s, several new or previously declining types of sexually transmitted diseases (STDs) had become wildfire epidemics. Genital herpes grew from a relatively minor health problem to a national concern. Unlike love, the grim joke says, herpes is forever. This painful, lifelong disease can not only spread to one's sex partners but fatally infect infants as they pass through the birth canal. Herpes increased more than tenfold from the mid-1960s to the mid-1980s; many of its sufferers formed support groups and took oaths of near chastity. They would have been even more frightened had they known that researchers suspected a link between herpes virus and cervical cancer. It turned out that the microbial villain in such cancer is not herpes virus but human papilloma virus (HPV), the cause of genital warts. Like herpes virus, HPV has become rampant. It was only the first of several increasingly widespread viruses found to play roles in common types of cancer.

When the AIDS syndrome was identified in 1981, it overshadowed all other STDs. It is a mass killer without parallel. Only rabies matches its lethal power, but rabies cannot be transmitted from person to person. Even in the best possible scenario—safer sex behavior, effective drugs and vaccines, and constant medical vigilance—AIDS will kill tens of millions of people in the next few decades, leaving entire nations invalids. It is not widely appreciated that viral hepatitis, often transmitted in the same ways as AIDS, sickens and kills even more people around the world each year, and continues to increase at an alarming rate.

As STDs changed and spread, so did more hemorrhagic fevers. A new child killer, dengue shock syndrome, has traveled from Asia to the entire tropical and subtropical world; mosquitoes capable of spreading it have entered the United States. The Seoul hantavirus, which causes Korean hemorrhagic fever, seemed a local problem when it attacked American soldiers in 1950. Then the virus was transported by ship to ports around the world. In 1985, a variety was found in Baltimore harbor rats, and then in Baltimore hospital patients with histories of stroke and kidney disease. Seoul-type viruses are suspected of causing such illnesses everywhere. One of their close cousins, the Sin Nombre virus that struck in the Four Corners region of the Southwest in 1993, is also more widespread than was first believed.

Where do these new diseases come from? So far, the only widely publicized answers are those of a few quixotic scientists who point to outer space, and of creationists who blame a vengeful deity. The claims for outer space have a deservedly small following; even some who proffer them do so with tongue in cheek. And many creationists claim just one miracle, the one they believe started the world's clockwork ticking. Some call AIDS a divine chastisement. So far, at least, they have not similarly blamed Lassa fever, Lyme disease, and legionellosis on the sins of Nigerians, suburbanites, and aging veterans.

New diseases do not fall from the sky or leap from some mysterious black box. Parasitism and disease are a natural, in fact necessary, part of life. They are basic to the existence of everything from the earliest, simplest organisms to humans. Diseases, old and new, strike horses, insects, plants, even bacteria. New ones are always coming into existence, most change with time, and some vanish from the earth. A small number of human diseases have always been with us, inherited from our primate ancestors. Chicken pox, for instance, struck the earliest humans, and it remains among us. But most human diseases were once new. They came to us because we changed our environment, our behavior, or both. Sometimes, as is happening now, they came in waves.

Most of these diseases came from other species—smallpox probably from dogs or cattle, hemorrhagic fevers from rodents and monkeys, tuberculosis from cattle and birds, the common cold from horses, AIDS probably from African monkeys. The vehicles by which many reached us were mosquitoes, ticks, and other small creatures that respond quickly to even minimal changes in our shared environment. Some of those changes happened naturally; at least as many we created.

We provide new ecological niches for microbes by tilling fields and domesticating animals, and by bringing into existence gardens and second-growth forests, villages and cities, homes and factories. We give them new homes in discarded truck tires and water tanks, in air conditioners and hospital equipment. We transport them by automobile, ship, and airplane. We alter their opportunities and affect their evolution when we change our abodes, our sex behavior, our diets, our clothing. The faster we change ourselves and our surroundings, the faster new infections reach us. In the past century we have changed the biosphere as much as any glacial surge or meteor impact ever has. So we and microbes are dancing faster than ever in order to survive each other. As we do so, the burdens on our environment and our immune defenses increase.

There is cause for alarm, but not for despair. Our primate ancestors had to cope with new diseases, and so did our Stone Age forebears. So did the first farmers and the first city dwellers. Despite struggles and crises, they were able to survive the challenges. And so, presumably, are we. The human immune system and human imagination are marvels of adaptiveness.

We are in one of those recurring eras of crisis when we accelerate the process of acquiring and adjusting to new pathogens. They, like us, are trying to adapt and survive. Some must be conquered; some require only a wise truce. If we are to adapt and survive, we must start by understanding how we have always coped with new diseases.

the first shocks

Origins. Disease is as old as life.
Eat or be eaten. The germ's point of view.
Symbiosis: a shared table.
Down from the trees and into the meat.
Is "Neanderthal" a compliment?

Living in rich, technologically advanced nations, we can imagine that we are more or less protected from nature, almost separate from it. Such illusion makes freedom from infection seem normal, health nearly a right. Illness takes the guise of an alien assailant. Like an impudent housebreaker, it makes us indignant, or perhaps we wonder what we did to give it an opportunity. If germs behaved properly, they would respect our intelligence and the usual boundaries of life. We could pursue our affairs without interruption.

That view is arrogant and dangerously misleading. Parasitism and infection are basic facts of nature, and new diseases have been evolving almost as long as life itself. As Sir Macfarlane Burnet and David White say in their classic *Natural History of Infectious Diseases*:

In studying the nature of disease the whole range of living beings comes into our province, for there is probably no species of organisms which has not at some time been either host to a parasite or a parasite itself. Many have filled both roles. Infectious disease is universal, and any attempt to imagine how it arose . . . will inevitably take us back to the very earliest phases of life.

The earth is some 4.5 billion years old. The oldest fossils, a billion years younger, are those of primitive bacteria; they resemble types recently discovered near hot mineral springs and seabed volcanic vents. The first life probably arose in the similar environment of hot primordial oceans. Perhaps lightning or ultraviolet light helped convert compounds there into amino acids, the building blocks of proteins, which are the basis of all life. Some molecules in this biological soup were able to duplicate themselves; they grew in complexity until they resembled DNA, the double-stranded substance of genes. Some of these proto-life forms developed envelopes that could contain an oceanlike microenvironment and its chemical reactions. The result was a primitive cell.

That, in simplified form, is the sort of description favored by most scientists since the 1950s, when some steps in the process were first duplicated in laboratories. Now we know that RNA, the single-stranded material that helps DNA produce proteins, can copy itself. Some scientists believe that before DNA appeared, there was an "RNA world" of early life forms. In either case, the earliest cells had no organelles, or specialized internal structures. As cells became more complex, they developed nuclei, where genes aggregate, and mitochondria, small bodies that produce energy and help regulate metabolism.

Infection was already ubiquitous when higher organisms left their first fossil traces, some half a billion years ago. There are fossil plants with fossil fungus infections, and ancient jellyfish and mollusks bearing signs of parasites. Dinosaur bones 250 million years old have marks of bacterial infection, as do the remains of mastodons and saber-toothed tigers. Today virtually every organism teems with smaller fellow travelers. Even free-living bacteria in the sea are infested with viruses. It is only when the diphtheria bacterium is infected by a virus that it produces toxin and causes human illness.

An increasingly popular theory puts parasitism at the very origin of complex cells. Early in this century, Russian biologist Konstantin

Merezhkovsky claimed that chloroplasts, plant cells' equivalents of mitochondria, began as bacterial invaders. In the 1920s, American biologist Ivan Wallin claimed that mitochondria, too, are survivals of ancient infection. Since the 1960s, biologist Lynn Margulis has elaborated the theory that all complex cells evolved through the merger of simpler ones, in what began as parasitism and ended in symbiosis.

For a while, these ideas lay at the fringes of scientific thought, but evidence has accumulated to support them. Mitochondria have their own DNA; they grow and divide on their own timetable, distinct from that of the nucleus. Cellular invasions can be witnessed today that resemble those theorized for the past; some simple cells enter nucleated cells and live cooperatively within them. And if the protozoon *Euglena* is treated with the antibiotic streptomycin, it is "cured" of its chloroplasts, and thus of its capacity for photosynthesis. Most biologists now agree that mitochondria began as primitive bacterial invaders that avoided being digested and were integrated by their hosts. Margulis says that such cooperation, or symbiosis, is the driving force behind evolution. Every organelle, she argues, started as an infection, and each human cell is a community of onetime invaders—in some cells, as many as eighty.

Other theories suggest a sort of backward evolution. One belief is that viruses are degenerated bacteria; they shed every structure save those needed to survive inside a cell (a virus, unlike most bacteria, can survive only within a host cell). Another theory holds that viruses evolved from cellular organelles, perhaps from mitochondria that escaped to a semi-independent existence. Both ideas have been disputed, but there is wide agreement on the bacterial origin of several types of tiny microbes (chlamydia, rickettsia, and mycoplasma) that were once thought intermediate between viruses and bacteria, and which cause human diseases such as trachoma and pneumonia.

These are only some of the theories in circulation. There are enough schemas of the origins of life and of parasitism to glaze the eyes of nonspecialists. They became increasingly intricate and varied with the rapid progress of molecular biology and its dazzling new techniques. Some bring to mind the definition of prehistory as "the study of the unverifiable to prove the unwarrantable about what never happened anyway." A few eminent scientists have even given a half-ironic nod to the "panspermia" theory, that influenza, AIDS, and life itself arrived from outer space, by comet or spaceship—a feat no more miraculous, they say, than the emergence of the simplest cell from a lifeless sea. Panspermia aside, most current theories show a trend away from the view of normal life as freedom from infection. Disease is not just biological thuggery, in which one species molests another. Rather, infection is an ancient event, basic to life, and it tends to lead toward peaceful coexistence.

Not all parasitism, of course, is benign. Directly or indirectly, every creature survives at some expense to others. It stays alive only if it creates proteins; to do so, it must take in proteins or the amino acids from which proteins are built. The ways one creature makes another's protein its own range from predation to parasitism, but all are paths to the same end. Viruses thrive on bacteria, trees, humans, and almost anything else alive. Fungi and bacteria scavenge plants and animals. Deer, horses, gorillas, and other munchers and browsers consume plants; carnivores prey on other animals. Multicelled parasites such as tapeworms attack their living larders from within, while many insects, ticks, and bats tap theirs from without. This endless protein transfer is sometimes called a food chain, yet in any biological arena, whether suburban backyard or tropical forest, it is more like a dense web of countless thousands of species and compounds.

The universal law "Eat or be eaten" may first evoke Tennyson's nature red in tooth and claw, but nature is not consistently ferocious. Predation and parasitism are self-limiting. If predators gobble up all available prey, they will eventually starve and disappear. Similarly, if parasites kill their hosts, they lose their meals, their homes, and thus their own lives. The ultimate adjustment between host and parasite is not murder but mutuality. Disease is a trauma that both, with luck, will survive. Fatal or severe disease is usually a sign that host and parasite are relatively new acquaintances. That is, the parasite has until recently been more at home in other hosts.

This is dramatically clear when one takes a germ's view of its encounter with a human. Perhaps the microbe normally dwells peaceably in a bird, squirrel, mosquito, or other species that has hosted the germ's ancestors for millions of years. Through casual contact— touching a squirrel, drinking water contaminated by bird droppings, attracting a mosquito that cannot find its usual meal of deer's or horse's blood—the microbe is transferred to a person. The germ plunges into a strange environment full of new opportunity and risk. It may die because of the new host's body temperature, the acidity of its tissues, or its arsenal of defenses, from engulfing white blood cells to the chemical weapons of its immune system. In a person who has previously met similar germs, these defenses have been honed to fine precision, to neutralize or kill the invader. The fate of most microbes that invade us is quick death.

Sometimes, though, a microbe finds in humans an ideal environment, with plenty of nourishment and mere Maginot resistance. And it has its own weapons; it can attack or elude white blood cells, produce toxins, and kill and feed on tissues anywhere from the toes to the depths of the brain. If it multiplies unhindered, it may kill the host. But should that happen before the germ finds transport to another home, the meeting becomes a dead end for host and parasite alike.

Even if the germ is successful, the host may mount an imperfect but adequate defense. The result is a passing illness, during which the germ's descendants reach new human homes (often by causing such symptoms as coughing, sneezing, or diarrhea). If the progeny of microbe and human remain in contact long enough, they tend to arrive at a standoff between their offensive and defensive weapons. For the microbe, this usually means adapting to a particular tissue or organ.

Such partnerships have developed at every level of the plant and animal kingdoms. There are "defective" viruses that can multiply only in cells that also contain "helper" viruses. Bacteria on the roots of legumes "fix" atmospheric nitrogen for the plants' use. Bacteria and protozoa in the termite's intestine feed on cellulose, breaking down the cell walls of wood the insect has eaten, thus releasing nutrients. Cows and elephants manage to thrive on bulky, low-protein diets only because bacteria ferment the meals in their stomachs. Luminescent microbes in the gut of the leiognathid fish lend light by which it hunts, confuses predators, and attracts mates. And in humans, the bacterium E. coli aids bowel function and deprives harmful germs of a berth. However, E. coli has adapted specifically to conditions in the colon. If it strays into the urinary tract, it causes cystitis. If swallowed in polluted water, it affects the upper gut and can cause dysentery.

The adaptation of parasite and host goes through stages called epidemic, endemic, and symbiotic. A germ entering a virgin population—one that is unfamiliar and has few defenses against it—often causes acute disease in people of all ages. This is the classic picture of an epidemic; if it involves much of the world, it is called a pandemic. The survivors are usually left with improved defenses against reinfection; over generations, additional defenses may develop. The disease eventually becomes endemic, a widespread, lower-grade infection or routine childhood disease. With further adaptation by parasite and host comes symbiosis, in which germ and host sustain mutual tolerance (mutualism) or even mutual benefit (commensalism, which literally means dining at the same table). Parasites and their hosts have always gone through these stages, and they still do so. Some biologists have argued recently that there is no trend from severe illness to symbiosis, and that virulence depends primarily on how a disease is transmitted. Transmission may indeed play a role in virulence, but the majority scientific view is that it is not paramount.

Infectious disease, then, is not nature's tantrum against humanity. Often it is an argument in what becomes a long marriage. This has been confirmed through laboratory experiment. Biologist Kwang Jeon infected amoebae with bacteria and bred the survivors; five years later, the amoebae could not grow without the bacteria. Outside laboratories, such adaptations may take hundreds or even millions of years. When germ and human meet, the initial advantage is sometimes the germ's, for it evolves through millions of generations during one human life span. However, human and microbial solutions to their respective challenges are many and ingenious. At stake for both are health and even life itself.

Fortunately for people, acute illness is more the exception than the rule in encounters with germs. There are hundreds of thousands of species of microbes; a pinch of soil holds millions of viruses, fungi, protozoa, bacteria, and other potential parasites. Few enter humans, fewer survive, and fewer still are passed to other humans. To cause human disease, a germ must beat enormous odds. If it does so, the reason is usually a change in the environment, in the germ's behavior, or in ours.

Such change probably has happened seldom in the history of most higher species, at times of ecological disruption—for instance, when an altered climate made predators seek new prey, or when hosts wandered into new regions and met unfamiliar germs. Except during a few such dramatic periods, our primate and hominid ancestors probably suffered as many diseases, new and old, as did other mammals. This is, they contended mostly with familiar, nonlethal infections and with contaminated wounds.

Whether these were the same microbes that exist today is not certain, but fossils speak eloquently of ancient trauma and infection. There are signs of fracture, infection, and healing in the bones of dinosaurs, mastodons, and early humans. Modern wound infections are usually caused by streptococci, staphylococci, or the germs of tetanus or gangrene. The latter two form spores that can survive in soil despite drought, heat, and cold; they still infect many wounds people incur out of doors. We cannot be sure that exactly the same germs afflicted dinosaurs and Neanderthalers, but even if the germs were not identical, the process must have been similar.

Some chronic infections, such as those causing dental decay and pyorrhea, have also left ancient stigmata. Jaws that ached from caries and abscesses existed from the earliest mammals to our nearest human ancestors. Also common in ancient bones is arthritis, which can result not only from wear and aging but from chronic infection. Severe arthritis afflicted prehistoric crocodiles, camels, bison, lions, and especially cave bears. In fact, it was so frequent in cave bears that the great nineteenth-century pathologist Rudolf Virchow gave it a special name, cave gout. Similar arthritis deformed the bones of Neanderthalers 50,000 years ago.

While many local infections leave distinctive marks on bones, few systemic ones do; even when fossil lesions exist, distinguishing one illness from another can be tricky. Certain changes in the bones of cave bears suggest that they had tuberculosis or a related disease, brucellosis. Epidemics that left no traces must have struck ancient species sometimes and even made some of them extinct. For now, the fossil record yields only tantalizing hints, such as the tsetse flies found in Colorado, dating from a million or more years ago. The tsetse, now limited to Africa, carries sleeping sickness, a protozoan disease that can lay humans low and wipe out vast populations of hoofed animals. The tsetse's presence in ancient North America corresponds roughly with an otherwise mysterious extinction of horses there.

Epidemics must have attacked our primate ancestors, just as they still strike our simian relatives. In this century, yellow fever imported from Africa almost eradicated howler monkeys in South America. Free-living primates suffer some of the same chronic diseases that afflict humans, and which presumably attacked hominids—malaria, hepatitis, yaws, tuberculosis, herpes simplex infection (cold sores), and invasion by such intestinal parasites as the pinworm, whipworm, and tapeworm.

Obviously, wild primates were not and are not creatures of perfect health, living in an uninfected Eden. Neither are humans, nor were they ever. Monkeys and apes inherited some diseases from their mammalian ancestors, and we in turn have inherited ills from those primates. Other infections, however, are relatively new, and they reflect changes, even crises, in the course of human history.

The first big shock to influence human disease patterns was our ancestors' descent from the trees to the ground, about five million years ago. Perhaps this happened when subtropical Africa became more arid, and savannas replaced forests. It has even been suggested that viral epidemics resembling polio or meningitis left our arboreal ancestors too crippled to swing through the branches, and enough survivors squeaked out a marginal adaptation to the forest floor to launch a new species. Whatever the reason for our ancestors' descent from the trees, it dictated shifts in their diet, lifestyle, and burden of disease.

As a species with our feet now firmly on the ground, we tend to think of territory horizontally. But every environment has significantly different vertical zones. In a forest, certain species of mammals, birds, and insects require the sunlight and food in the leafy canopy; others need the shade, moisture, and food on the ground; several intermediate zones may exist between earth and treetops. Changing its niche by only a few meters can radically alter a species' prey, predators, and microbes.

Today we often see diseases invade new vertical zones. In Central and South America, mosquitoes infect treetop monkeys with the yellow fever virus. The disease remains isolated in the canopy because monkeys and mosquitoes there rarely travel lower. The commercial demand for tropical timber has sent loggers into the forests, and when they fell a tree, clouds of mosquitoes come to earth with it. The mosquitoes then feed on the primates nearest at hand, the ones with the axes, and transmit the virus. On returning home to cities, infected lumberjacks set off urban epidemics of yellow fever.

Our ancestors' descent to the ground freed them from some old diseases but exposed them to new ones, through ground-level air, water, and foods. They probably acquired parasitic worms from the flesh and droppings of savanna herbivores; sleeping sickness was transferred from those herds by tsetse flies. Further changes occurred when the early hominids called Australopithecines split into two evolutionary branches. The so-called robust type had big teeth and heavy jaws, for eating plants, seeds, and nuts; they eventually died out. The more slender, or gracile, from whom we are directly descended, became omnivores and eventually hunters.

So far we have only poignant hints of what these creatures were like. There are some smallish, fragmentary skeletons and, at Laetoli, in Tanzania, a little group's footprints left in fresh volcanic ash and miraculously preserved three and a half million years later. We may never know in what order they developed distinctively human traits—fully upright posture, bipedalism, an enlarged forebrain, prolonged childhood, the use of fire and shaped tools, group hunting, speech, and complex social organization. The appearance of humans was not an event but a process; perhaps it occurred more than once, and differently in different places.

We do know that close to two million years ago, Australopithecines were replaced by *Homo erectus,* our first fully upright, large-brained ancestor. He learned to use fire and make stone tools. He rather resembled modern *Homo sapiens,* but he would stand out today in one's living room. He probably was not capable of speech, let alone social chatter. Still, his changing way of life brought the next major shift in the human disease pattern. The biggest reason was meat.

Now we enter grounds of contention. The behavior of *Homo erectus* has provoked truly savage acrimony among scientists. Was our ancestor a killer ape or a gentle forager? Was he more predator or prey? Were his first tools weapons? Was the growth of his intelligence driven by aggression and hunting? These are really arguments about human nature, and they stir the sort of anger usually reserved for debates on race and sex. Fortunately for *Homo erectus,* he is safe from modern academic rage.

Each year more fossils appear, and they are studied in ways that until recently were unimaginable. Fossil teeth are examined with the electron scanning microscope to distinguish the wear marks of meat and of plant foods. Laboratory analysis reveals the isotopic signatures left in ancient bones by dietary plant and animal proteins. In the light of such knowledge, only the most ideologically partisan researchers deny that hominids became scavenging omnivores and then group hunters. They came to obtain one-quarter to one-half of their protein from meat.

This change probably took tens or hundreds of thousands of years, and it required shifts in genetically determined aspects of body and behavior. Pathologist Michael Zimmerman, studying early *Homo erectus* remains from Kenya, found changes that look like the result of toxic megadoses of vitamin A. Zimmerman thinks this may reflect a diet rich in the internal organs of animal prey during "a somewhat experimental meat eating phase of human evolution."

The effects of adding hunters' feasts to hominids' predominantly plant diet were vast. As one learns by observing modern herbivores and carnivores, meat means time. A horse or monkey, which must process so much low-protein herbage, spends much of its life eating. A carnivore, thanks to its meals of concentrated protein, may feed only once every few days. For early humans, with their large brains and manual deftness, the time freed by eating meat could be spent making tools to catch more of it, and for other purposes as well. There was time to plan hunting in groups, to pray for meat, to sing in celebration of it, to depict animals that provided it. This is, there was time for creating culture.

A flexible, omnivorous diet enabled our precursors to survive where plant food was scarce but game was plentiful. With spears that could kill big prey, they expanded into temperate climates, then to near deserts and tundra. Humans became the only primates to inhabit the entire world, from equatorial jungles to subpolar regions.

The dispersal began a million years ago, when *Homo erectus* fanned out from Africa to the warmer parts of Europe and Asia. In doing so, he left behind a host of tropical parasites. In warm, moist climates, many worms and protozoa can survive outside their hosts, and unlike bacteria and viruses, they tend to provoke little or no immune reaction. Therefore they can keep reinfecting individuals and communities with chronic diseases. Multiple parasitisms still hinder development in many tropical regions. When *Homo erectus* left home, he abandoned such companions as the worms that cause river blindness and elephantiasis. But for each disease he escaped, he met a new and often worse one.

This change in disease burden may have exceeded that caused by leaving the trees. In each new ecosystem, nomadic hunters met new prey, new vectors (disease carriers), and new parasites. The result was an onslaught of zoonoses, animal infections that can be transmitted to humans. Being new to people, the germs often caused far worse symptoms than in their usual hosts. Such virulence is so predictable that any deadly human infection should be suspected of being a relatively recent resident of our species.

We will probably never reconstruct in detail the waves of zoonoses that struck humans in their global dispersal, but recent centuries, even recent decades, hold examples of people catching unfamiliar illnesses from new prey. For instance, the crew of the Danish explorer ship *Unicorn* repeated what must have been a common disaster for *Homo erectus,* a story that became clear some 350 years after the fact.

In 1619, the *Unicorn* sailed in search of the Northwest Passage, the fabled water route across North America between the Atlantic and Pacific oceans. The crew might have been forgotten, like many others who tried and failed, had it not been for their peculiar end. Sixty-one of the sixty-four crewmen suffered an agonizing death in Canada's frozen Hudson Bay in 1620; most were buried in unmarked graves near Churchill, in Manitoba. Historians long repeated the guess that they had died of scurvy. That disease, caused by severe vitamin C deficiency, killed countless seamen until the late eighteenth century, so the guess seemed plausible.

In the 1970s, Canadian historical writer Delbert Young became suspicious of the diagnosis. He learned from the memoir of the *Unicorn*'s captain that the crew became sick after eating raw polar bear meat (not for lack of fire, but because Europeans considered raw bear meat a delicacy). Infected bear meat, Young knew, had recently been identified as the killer of Swedish explorers who in 1897 had tried to reach the North Pole by balloon; decades later, their frozen supplies were discovered, studied by a Danish physician, E. A. Tryde, and found to contain *Trichinella* parasites.

Today we think of potentially deadly trichinosis as caused by contaminated, undercooked pork. Once widespread in the United States, it became uncommon and then returned in the 1970s, mostly among Southeast Asian refugees who ate pork raw or lightly cooked. Trichinella, though, is not only a pork parasite. The worm and its cysts infect such carnivorous mammals as bears, foxes, and walruses, especially in cold climates.

That knowledge, and Tryde's discovery, prompted Delbert Young to scrutinize the records of the *Unicorn* and other northern expeditions. He concluded that trichinosis had ended the *Unicorn's* voyage and that of an English expedition that had preceded it to Hudson Bay a decade earlier. Just as malaria, yellow fever, and sleeping sickness slowed first the settlement and then the European exploration of West Africa, such prey infections as trichinosis may have delayed the human occupation and exploration of the world's subpolar regions.

In our own century, contacts with prey species revived the deadliest epidemic disease in human history, bubonic plague. In the 1970s, a small resurgence of plague began in the United States, mostly in the West and Southwest. In 1983, there were 40 cases and 6 deaths, the highest toll in sixty years. New cases still appear each year. Many varieties of rats, mice, ground squirrels, and other wild rodents have lived with the plague bacillus for millions of years, but to humans it is a relatively new and lethal acquaintance. During the 1980s, one case after another was traced to such incidents as someone's skinning a prairie dog, handling a wild chipmunk, or petting a cat that had caught a wild rabbit.

Such cases suggest how *Homo erectus* first fell victim to bubonic plague. Microbiologist Charles Gregg, an expert on plague, writes: "The first human case may have occurred when the earliest hominids began to vary their diet by running down small game. A sick animal is more easily captured than a healthy one, but the captor risks taking disease as well as sustenance from his prey." Plague probably wasn't epidemic until a couple of thousand years ago, when people lived in large settlements, and a disturbed ecosystem caused unusual migrations by rodents. But plague, like trichinosis, must have occasionally attacked individuals and small bands of humans before that.

Thus *Homo erectus* gradually added to his accustomed ills, mostly chronic and passed from person to person, a formidable list of zoonoses. Scavenged meat might carry botulism or staphylococcus infection. Butchering a kill could cause gangrene or tetanus; skinning it could bring exposure to tularemia or anthrax. Wild game transmitted the microbes of relapsing fever, hemorrhagic fevers, brucellosis, leptospirosis, toxoplasmosis, and salmonellosis. Unfamiliar insects and ticks carried scores of diseases, such as scrub typhus and encephalitis. And if eating moose or mammoth did not cause trichinosis, it might transmit tapeworm or other debilitating parasites that cause vulnerability to more severe diseases.

If we consider the number and severity of such zoonoses, it may seem amazing that *Homo erectus* survived and thrived. But certain aspects of nomadic hunter-gatherer life worked in his favor, especially outside the tropics. A few unfamiliar germs, such as those of typhus and sleeping sickness, adapted to humans as secondary hosts, but most failed to find a dependable means of travel from person to person. These diseases remained sporadic, rising from occasional human contact with animal hosts. And parasites, like most kinds of organisms, are less varied and numerous in cool climates than in hot ones. Since more nontropical infections are caused by bacteria and viruses than by protozoa and worms, there is a greater chance that survivors of nontropical diseases will be immune to reinfection.

What most protected early humans from being wiped out by new diseases were their small numbers and frequent movement. We infer this not only from the fossil record but also from detailed studies of present-day hunter-gatherers, such as Australian aborigines and the Bushmen of the Kalahari Desert. They have, of course, undergone physical and cultural evolution since the Paleolithic (Old Stone Age); they are not, as was once thought, "living fossils." They also have had contact with some of the diseases of outsiders. But they do offer a diffracted image of life in the past.

Ancient hunter-gatherers lived in bands of probably a few dozen, perhaps a hundred at most. They roamed fairly large areas, seldom exceeding a density of one person per square mile. Unable to preserve and store food, they had to move and seek new sources often, but they went no farther or faster than their feet could carry them. They did not live with heaps of garbage and feces or with polluted water. They rarely suffered the shock of drastic environmental change.

Nomadic bands may have had occasional contact, even some large gatherings for trade or rituals, but populations were far too small for crowd diseases to take hold. There was little or no flu, measles, mumps, whooping cough, typhoid, or smallpox. *Homo erectus* may have come to consider sporadic attacks of plague or trichinosis normal, but otherwise, like many modern hunter-gatherers, he was probably healthy and well nourished, with an average life span close to forty years. That is, *Homo erectus* was physically much better off than billions of people in the Third World today.

So when Neanderthalers appeared, more than 100,000 years ago, the human evolutionary line had already experienced three great changes in disease burden. One was caused by the ancestral descent to the ground, another by humans' becoming omnivores and hunters, the third by their entering new environments. Before Neanderthalers vanished, around 35,000 years ago, they left behind one of history's most revealing patients, a battered, aging man his discoverer nicknamed Nandy. His life and death offer a revelation not about human disease but about sustaining human life.

When Nandy was found, in the 1950s, Neanderthalers were not in high esteem. The first one had come to light a century earlier in Germany's Neander valley, near Düsseldorf. Big, stocky, and heavily muscled, with powerful jaws and a forward-projecting face, he was not quite modern *Homo sapiens.* Although his brain was as large as ours, perhaps a bit larger, his stone tools impressed scientists, then reveling in high Victorian culture and the miracle of the steam engine, as woefully crude.

To generations giddy with the idea of evolutionary progress, the Neanderthaler embodied everything primitive—greed, lust, violence, and dim wits. Early in this century, French paleontologist Pierre Boule fortified a Hobbesian image of Neanderthal life as nasty, brutish, and short. Misreading arthritic spinal changes, he declared that the remains of the famous Neanderthal "Old Man of La Chapelle-aux-Saints" had a naturally stooped posture. Scots anthropologist John McLennan had already invented a vision, still perpetuated in cartoons, of men in fur suits clubbing women and dragging them by their hair to bridal lairs. It is no wonder that museum dioramas and children's books have portrayed "cavemen" such as Nandy hulking about pathetic little fires, stooped and less expressive than oxen. Hollywood shows them snarling and bashing each other over meat and mates. To this day, the word "Neanderthal" is less than a compliment.

This picture is in some ways misleading, in others clearly wrong. Neanderthalers were probably the first humans to live regularly in cold climates; to achieve this, they created clothes, shelter, and new tools. They buried their dead, often in family groups, with objects suggesting that they believed in an afterlife. Compared with recently discovered types of hominids and earlier humans, the Neanderthaler seems not primitive but rather advanced. Still, a major revision of his portrait came only after the work of anthropologist Ralph Solecki.

In Shanidar cave, in northern Iraq, Solecki discovered the 60,000-year-old bones of nine Neanderthalers. The first skeleton he unearthed is referred to in scientific literature as Shanidar 1, but Solecki came to feel a personal closeness to him and bestowed the nickname Nandy. Nandy's life was hard, but neither very short nor unrelievedly brutish. In fact, his broken bones testify to a collective triumph over hardship and suffering.

Years before Nandy died, at age forty or more, he had already survived multiple fractures of the skull, legs, hands, and feet, and a crushing fracture over the left eye that deformed his face. The wounds may have left him partly paralyzed on his right side and blind in his left eye. His right arm was gone just below the elbow, perhaps amputated after a crushing injury. The pattern of wear on his teeth implies that he grasped objects with his jaws and manipulated them with his left hand.

Probably some of Nandy's wounds were caused by fights with stone weapons, others by rockfalls or hunting accidents. Whatever their source, they left him barely able to forage, and surely unable to survive on his own. He could have reached middle age only if he received rest, feeding, and long, intensive care, not once but several times in his life. His entire band must have protected him and shared their resources with him while he could contribute nothing tangible. After he finally died in a rockfall from the cave's roof, stones were carefully piled on top of his body. The bones of small mammals upon those stones may be the remains of a death feast.

The other human bones in Shanidar cave show that Nandy's case was not unique. The skeleton called Shanidar 3 was that of a man who died at age forty to fifty, having survived by many weeks a penetrating chest wound that marked one of his ribs. He probably was recuperating when killed in a rock fall. His right foot bears signs of degenerative arthritis severe enough to cause a painful limp. Like Nandy, he must have received help and care.

Shanidar cave held more surprises. When paleobotanist Arlette Leroi-Gourhan analyzed soil from the grave of Shanidar 4, she found fossilized pollen from brightly colored wild flowers. There was no way it could have reached the cave burial by chance; the dead man's kin must have gathered a floral tribute for his grave. With the discovery of those flowers, a strong, intimate emotion flashed across a silence of 60,000 years, and it convinced Solecki that Neanderthalers were not distant, dimwitted cousins. "In Neanderthal man," he wrote, "we recognize the first stirrings of the concept of man caring for his own, a sense of belonging and family."

An even greater surprise came when researchers realized that most of the flowers in the grave, such as hollyhock and yarrow, have medicinal powers. They are used today in folk medicine, for treating ills from toothache to wound infections. Neanderthalers fought injury and disease not only with physical care but with healing lore. It should be no surprise that such people were no longer merely being shaped by their environment; they were actively responding to it. Soon they would start to change it drastically.

Three

revolutions

The age of overkill. Healthy nomads;
farmers were shorter, so were their lives.
The perils of plenty. Taming the beasts
and the fields. More zoonoses.
The Mystery Disease of Pudoc.

The terms "Neolithic Revolution" and "Agricultural Revolution" describe technology that 10,000 years ago led first to village and then to urban life. Both suggest a triumphant leap from nomadic darkness to sedentary light. On an evolutionary time scale, the changes were swift, but in terms of human experience, there was no quick leap; rather, a process long under way gathered speed. It brought more food, swelling populations, and humanity's worst wave of new diseases—what we recognize today as the distinctive human pattern of infections.

Actually, the Agricultural Revolution was both a triumph and a disaster. Humans became the victims of plenty, falling prey to scores of diseases they had never known. The two changes, toward greater plenty and more infections, were inseparable, and they still occur together; as we shall see with the Mystery Disease of Pudoc. But first we must see just how the Neolithic (New Stone Age) Revolution arrived.

The word "revolution" is often abused today. This reflects our era's predilection for seeing change rather than continuity. Hardly a leaf falls or a placard is raised without someone's declaring another revolution— technological, social, sexual, or merely journalistic. Thus the drama of real revolution is trivialized. However, the idea of the Neolithic as a revolution is legitimate. One must imagine the astonishment when archaeologists in the first half of this century unearthed proof of what then seemed civilization's stunning antiquity. (By civilization, they meant urban life based on extensive, systematic farming.) Older scientists could recall when the very ideas of evolution and a long prehistory had been shocking. They and their teachers had had to argue with the computations of the learned and defunct Ussher.

Archbishop James Ussher, like most intelligent seventeenth-century scholars, believed in divine creation. Applying his methodical labors to the Bible, he pinpointed God's creation of earth to the year 4004 B.C. Ussher was so painstaking and authoritative that the date was inserted in the margin of the King James Bible to enlighten readers of Genesis. Two centuries later, critics of Charles Darwin cited Ussher: Man could not be the product of so long and slow a process as evolution, since the world itself was only 6,000 years old. Mere millennia, they said, separated Darwin from Adam and Eve. And so William Jennings Bryan would still say in 1925, in a last-gasp defense of Ussher at the John Scopes "monkey trial."

The archbishop, a great scholar of his day, is now recalled as a paradigm of pious error. I raise his name not to mock him for having lived and thought in the wrong century, but to emphasize how amazing it was to learn that great cities, supported by vast farms and herds, had existed in the Tigris-Euphrates valley 6,000 years earlier. It would later be equally shocking to learn from such caves as Lascaux that earlier people had created art of such power and elegance, and that their culture and technology surpassed anything historians had envisioned.

Research has continued to enrich the panorama of human prehistory. Now we know that the Neolithic Revolution was preceded by a much longer "broad-spectrum revolution." That in turn was preceded by an important shift in subsistence technology that may have started as many as 50,000 years ago. To avoid dubbing that, too, a revolution, one might call it the "age of overkill."

This is a modest name for the first time people changed the environment more than it changed them. Group hunters all over the world ravaged their ecosystems by efficiently pursuing the biggest game. They rose to the top of the terrestrial food chain and triggered cascades of changes that brought a significant increase in human infections. Eventually, with the emergence of agriculture, they set off an epidemiological crisis.

This could have been done only by modern humans, *Homo sapiens*. They originated at some still undetermined time in Africa or Eurasia or both.* They spread to the Near East no later than 90,000 years ago, Australia 45,000 years after, and Western Europe 15,000 years after that. They reached the New World at least 12,000 years ago, perhaps much earlier. By 35,000 years ago, the Neanderthalers had died out. Modern humans, with the world to themselves, knew better than any of their extinct ancestors how to create livable microenvironments for their once semitropical species. They occupied every landmass in the world except Antarctica, and everywhere they went, big life forms dwindled as humans hunted in groups.

Mass extinctions of big species began to occur in Africa around 50,000 years ago, in Europe and Asia around 20,000 years ago, and in the Americas around 11,000 years ago. Scientists first blamed climatic changes; they could not believe that Stone Age hunters wiped out giant sloths, saber-toothed tigers, and mastodons. Some scientists maintain that view, but the evidence has become impressive that skilled, wasteful hunters armed with spears and other weapons could and did exterminate up to 90 percent of the larger species in one part of the world after another. Certainly people with similar technology achieved similar feats in historic times. Less than a thousand years ago, Polynesians with Stone Age tools became the first migrants to many Pacific islands, and in mere centuries they hunted huge numbers of species to extinction.

---

* I will not detail here the questionable theory of an "African Eve" or the competing idea of a Eurasian or multifocal origin of *Homo sapiens*. And I will avoid arguments about whether the New World was settled 40,000 or 12,000 years ago. Controversies on these and other, related chapters of human prehistory are shifting almost yearly now, as new evidence appears.

Hunting new game in new environments brought zoonoses, as it always had, but greater effects of big-game overkill were to follow. As the protein bonanza of big kills dwindled, people had to devise more efficient ways to feed themselves. They invented better traps and weapons for hunting small mammals and birds. They lingered on shores and riverbanks to feed on fish, mollusks, and crustaceans. With polished stone adzes they could fell trees and build boats from which to hunt marine prey with harpoons and nets. In the Near East, they learned to anticipate the migrations of hoofed animals and to stampede them into killing traps. In Australia, they hunted big and then small game with controlled brush fires. Everywhere they tried eating new seeds and vegetables, and invented new ways to process and store them.

This broad-spectrum revolution intensified around 15,000 years ago in the Old World and 8,000 years ago in the New. Excavated campsites of that era yield hearths, kilns, stone lamps, grindstones, awls, tally sticks, and sculpted human and animal figures. Pigments, metals, and shells were traded; perhaps, as among hunter-gatherers today, there were networks of exchange and loose federations of groups. Social evolution, not biological evolution, was now driving change in the human species, faster than it had ever been driven.

People were not yet fully sedentary, but they stayed longer at their camps. Populations probably grew; some settlements may have numbered as many as several hundred people. As they stayed longer in one place, they created middens of bones, seashells, garbage, and feces. These bred microbes and germ-carrying insects, and drew scavenging bird and rodents, all bearing new infections.

The knowledge that human diet and disease expanded together comes from a multitude of interlocking specialties. Physical anthropologists have found that plant and animal foods leave distinctive chemical signatures in human bones, as do proteins from marine and land animals. Paleoparasitologists, who prove that one man's mess is another's treasure, study coprolites, or preserved feces, for fossilized parasite eggs. Teams of archaeologists and paleobotanists sort through the garbage, ashes, and fossilized pollen at ancient campsites to learn what foods people ate. Paleopathologists examine bones and naturally preserved bodies to learn about ancient health and sickness. The result of all this is a record of declining robustness and increasing infections late in the broad-spectrum revolution.

Intensive studies have been made of the remains of the Palomans, who once lived near the Peruvian coast south of present-day Lima. Their bones show that by 7,000 years ago, they depended heavily on shellfish and other marine foods. Their coprolites and garbage reveal that they thus acquired the fish tapeworm and the lung fluke *Paragonimus*, which causes symptoms resembling tuberculosis, and which people still catch from eating infected crabs and shrimp. Perhaps because the Palomans relied on plentiful marine proteins rather than farming, their overall health and longevity were rather good. By 5,000 years ago, when they had turned to farming and a diet heavy in carbohydrates, their size and health had declined.

There are three good indicators of health in prehistoric remains; these are Harris lines, Wilson bands, and enamel hypoplasia. During the growth years, illness or malnutrition makes bone growth slow or even stop. Recovery brings a growth rebound that shows on X rays as distinctive marks called Harris lines. Their number and position tell how many insults to health a person has suffered, when they occurred, and sometimes how severe they were. The record is not perfect, for some lines fade, and they cannot reveal the exact reason for a stress. More reliable testimony to change are marks in the teeth, Wilson bands and enamel hypoplasia; both are caused by the slowing rather than the rebound of growth. Each of these three indicators has limited precision and completeness, but together they give a picture of health or disease in individuals and in populations.

These signs reveal that while the broad-spectrum revolution was a triumph of ingenuity, it meant diminishing returns in exploiting the environment. One food source after another was depleted and replaced. The process is clear at Tell Abu Hureyra, a great mound on the bank of the Euphrates, in northern Syria. It contains garbage, including charred grain and animal bones, that accumulated during thousands of years of habitation.

In the late Mesolithic (the transition from the Old to the New Stone Age), around 11,500 years ago, Abu Hureyra probably held a few hundred people. No longer nomads, they lived in wood-frame huts. The biggest game had been killed off long before, but people still hunted herds of wild gazelle and an occasional wild goat or pig. More often they killed rabbits and other small game, and ate fish and mussels from the Euphrates. Besides gathering fruits, nuts, lentils, and wild wheat and barley, they may have begun cultivating crops on a small scale and herding half-tame sheep and goats.

A 500-year break in occupation began around 10,000 years ago. When it ended, Abu Hureyra hosted a Neolithic culture. Now there were several thousand people living on fewer than thirty acres, many in permanent houses of mud brick. Grains and lentils were farmed intensively. (We know this because cultivated grain looks different from wild grain, just as domesticated pigs differ from their ancestors, the wild boars.) Sheep and goats were not hunted but herded. Soon gazelles almost vanished, and pigs and cattle were domesticated. Small-city life had begun; artifacts reflect trade with places as distant as Turkey and Sinai. This culture would flourish at Abu Hureyra for several thousand years. Then the Near East saw the final steps of Neolithic technology—fortified cities, plowing, irrigation, fertilizers, and the use of animals for power and travel.

Thus sedentism, farming, and animal husbandry came not in a transforming flash but gradually, and they coexisted with hunting-gathering for thousands of years. These developments came at different times and in different orders around the world. Full dependence on farming arrived in much of North America only a millennium ago. Wherever and whenever it came, it brought declines in health and an increase in diseases.

Old World skeletons from the late Mesolithic average two inches shorter than those of Paleolithic hunters. Their bones, whether examined by the naked eye or through a microscope, seem less robust. North American skeletons and teeth show the same decline during the comparable Archaic–Woodland transition, from early to late hunter-gatherer life. The gravesite proportions of children and adults suggest a rise in infant mortality.

On the eve of the Agricultural Revolution, people in much of India, the Middle East, Europe, and North America were already less healthy and less well nourished than their hunter ancestors. The trend continued in the Neolithic because of a vicious synergy of changing lifestyles, declining nutrition, and new infections. There were three kinds of new diseases—occupational, nutritional, and infectious.

Occupational ills arose because the greater size and complexity of societies brought specialized division of labor. A few million years of nomadism had not prepared the human body for the literally back-breaking work of full-time farming. Neolithic skeletons everywhere show a dramatic increase in arthritis and in stress fractures of the lower spine and load-bearing joints. As cities grew, so did the number of specialized crafts, each with its own type of physical punishments. We know this partly because the body remodels bone to accommodate the attachments of overused muscles. One can identify a javelin thrower's elbow, corn grinder's wrist, and scribe's finger in ancient skeletons, as one can detect tennis elbow or pitcher's shoulder in modern ones. By the age of metals, occupational ills would be even more varied and widespread; some, such as those related to tanning and smelting, involved toxic materials that probably predisposed people to infectious diseases.

Agriculture had brought a stable, year-round food supply that could expand along with the population, but it made nutritional ills common. A full belly did not guarantee a healthy body; dependence on starchy staples was quite destructive. Maize, rice, wheat, potatoes, yams, and manioc were high in carbohydrates and calories, but unlike meat and other plant foods, they were low in protein and certain vitamins and minerals. The result of such hollow plenty can be seen today in poor nations, and in poor pockets of wealthier ones, where calories are adequate but other nutrients are not.

Early villagers of Denmark and Norway, more than 2,000 years ago, had the knock knees and flattened pelvic bones of rickets, caused by lack of vitamin D. At the Paloma site in Peru, as elsewhere in the world, more carbohydrates meant more dental decay and root abscesses. Skeletons from the Dickson Mounds, in Illinois, show the price of over-dependence on maize. Between the years 950 and 1200, the proportion of children there with porotic hyperostosis more than doubled, from 14 percent to 32 percent. This spongy thickening of the bones results from the body's attempt to churn out extra red blood cells, to compensate for iron-deficiency anemia. Childhood anemia also soared in early farming settlements in the Near East and the Mediterranean. It struck especially at weaning age, when mother's milk was replaced by staple cereals.

The problem was not just lack of iron; poor nutrition is rarely that selective. People who lack iron may also lack the vitamin C needed to absorb it or the protein needed for hemoglobin synthesis. The resulting decline in health is complex and general. Bones at the Dickson Mounds show not only poor nutrition but a doubling of bacterial bone infections. We have seen that Mesolithic people were smaller and shorter-lived than Paleolithic hunters. Neolithic farmers were smaller still, and in places as different as ancient Japan and North America, their lives were at least several years shorter than those of their Mesolithic ancestors. As agriculture spread from its sources in the Near East, China, and Mexico, life expectancy in much of the Neolithic world dipped from the usual forty years of hunter-gathers to about thirty.

Poorly nourished people are an epidemic waiting to happen. By an exquisite misery of timing, the Neolithic vulnerability to disease came as people were exposed to a torrent of new pathogens. Their wastes, garbage, and granaries drew scavengers, and they were domesticating many species, from horses to chickens, which bore hundreds of unfamiliar parasites. Each creature in the human orbit was exposed to infection by the others. The result for all was a potentially disastrous biological stew.

Humans' first animal companions were dogs, bred from the wolves and jackals that scavenged at the edge of hunters' camps. Perhaps, as in Konrad Lorenz's reconstruction, domestication began when a little girl brought an orphaned pup to camp and persuaded the other people not to roast it. Deeply social, jackals and wolves could easily be raised to bond with humans. Intelligent, they could help men hunt, herd flocks, and act as alarms, companions, pets, and meals of last resort.

For the first time, people lived in continual intimate contact with another higher species. Man and dog worked and played together, fed and slept together, urinated and defecated in the same areas. Cheek by jowl, skin by fur, they inevitably exchanged pathogens. As a result, humans ran higher risks of rabies and became hosts to new types of worms, to tick-borne diseases such as tick typhus, and to echinococcosis, a nasty infection that causes cysts of the liver and lungs.

A similar rise in diseases came with each domesticated species. Birds and pigs, like dogs, probably drew close to humans by scavenging around their camps and villages (the chicken and domesticated pigeon are descended, respectively, from the wild rock dove and the Asian wild jungle fowl). Scavenging rats and mice attracted cats, which were domesticated as pets, rodent killers, or both. Small mammals such as rabbits and guinea pigs were captured and bred for food. By about 4,000 years ago, horses, oxen, goats, and sheep had been tamed and bred to provide food, labor, and transport.

Modern city dwellers must keep in mind that these animals, with their viruses, bacteria, and worms, did not live in some distant reserve or hygienic compound. Humans and livestock often lived under the same roof, on the same floor of dirt or straw. People caught the animals' germs by breathing the same air and dust, touching the animals' wastes, butchering their bodies, using their wool and hides, consuming their milk, eggs, or flesh. Many of the germs they met could not survive in humans, but others found a congenial second home. Cats, dogs, ducks, hens, mice, rats, and reptiles, for instance, can all carry *Salmonella* bacteria, which in humans cause mild to deadly intestinal infections. People have been catching salmonellosis from poultry, eggs, and fecally contaminated water for 10,000 years—most recently, in the United States, because of crowded conditions in poultry farms and speeded-up, automated processing of chickens and eggs.

Humans also acquired from livestock and pets a heavy load of helminths, or wormlike parasites. Such intruders range from the microscopic filarial worms that cause elephantiasis and African river blindness to yards-long intestinal hookworms and tapeworms. The effects ranged from moderate loss of blood and nutrients to disability and death. Parasitologist Michael Kliks estimates that a few million years ago, hominids carried perhaps a half-dozen such parasites. They were what J. F. A. Sprent calls "heirloom species," inherited from ancestral primates. To acquire more helminths, says Kliks, all humans had to do was to "bite into the apple of broader environmental and dietary exposure."

As hunters, humans had nibbled at the environment, occasionally catching from their prey such ills as anthrax and trichinosis. As farmers, they bit off almost too much to digest. There are now more than a hundred common human helminths, and many uncommon ones. Most are what Sprent termed "souvenir species," acquired by contact with prey and domesticated animals. The fact that no present-day genus of human helminth lives in humans alone shows that they adapted to us from homes in other creatures.

Over the millennia, intestinal helminths may have caused greater human damage than some of the more dramatic bacterial and viral plagues. Among the first victims were Neolithic farmers suffering the synergy of poor nutrition, intestinal parasites, and microbes. Dependence on starchy staple crops deprived the body from without, hookworm and tapeworm bled it from within; then bacterial and viral infections shortened a debilitated life. The skeletal signs of this toll— porotic hyperostosis and pitted orbital and cranial bones—can be seen in the remains of Neolithic children in Southeast Asia, Australia, Hawaii, Greece, and Costa Rica. Such interplay of helminths and dietary deficiency may have fatally eroded the Maya empire, as it enfeebles many of the Mayans' descendants today in Mexico and Guatemala.

Humans were not the only sufferers in crowded farms and villages. Bovine tuberculosis, an ancient disease of cattle, is rarely epidemic in the wild; domestication made it common and spread it to humans. Similarly, *Brucella* bacteria are common in wild ungulates but rarely cause epidemics. In farm herds, brucellosis can run rampant, causing sickness and spontaneous abortions. People catch it from cows as undulant fever, so called for victims' wavy fever charts. And the protozoon *Trypanosoma brucei*, which causes only a mild infection in wild African ungulates, strikes domesticated herds with deadly nagana disease, and humans with lethal sleeping sickness.

A number of diseases, such as malaria, yellow fever, and influenza, have been passed back and forth between humans and other species many times. The flu virus has an ancient home in birds and swine; since their domestication, birds and pigs have exchanged mutating and recombining flu viruses with humans, in variants for which no immunity exists. Periodically the result is a killer pandemic, such as the one of 1918.

Agriculture brought humans so many new pathogens that it seems wondrous they survived. Fortunately, the new diseases did not all appear and spread at once. Some remained occasional, dead-end incidents until human populations were big enough to sustain crowd transmission. Flu, smallpox, measles, and mumps probably began as sporadic zoonoses from domesticated animals. The measles germ is related to the viruses causing distemper in dogs, rinderpest in cattle, and a type of swine fever; any of these may have sparked the human disease, though the distemper virus seems the best candidate. The smallpox virus is kin to those causing vaccinia in cows, ectromelia in mice, and pox infections in fowl and swine. Such zoonoses ticked away in village and barnyard, biological bombs awaiting dense human populations.

Biologist Thomas Hull compiled a list of diseases people had acquired from domesticated animals, and grouped them according to their probable sources:

Dogs	65
Cattle	45
Sheep, goats	46
Pigs	42
Horses	35
Rats, mice	32
Poultry	26

This is not a complete, up-to-date list of zoonoses; it does not include such draft animals, food sources, and pets as camels, llamas, rabbits, guinea pigs, cats, monkeys, fish, and reptiles. Russian biologist Evgeny Pavlovksy estimated that we share almost 300 diseases with domesticated species and about another 100 with wild birds and animals. His estimate, like Hull's, may be low; more zoonoses have been discovered since those lists were made, and new ones are still coming into existence.

It was not domesticated animals alone that brought new diseases to Neolithic humans. Virtually every step our ancestors took to increase and vary their food supply invited novel infections. When Neolithic farmers cleared land for planting, pasture, or timber, they came in daily contact with species they previously had met only in passing, from monkeys to mosquitoes. The process continues today in much of the world. Americans in newly created suburbs have caught Lyme disease from deer ticks, and bubonic plague from the fleas of scavenging rodents. Africans clearing virgin land have caught hemorrhagic fevers from monkeys and wild rats. They may have caught AIDS from monkeys as well.

Clearing timber, slash-and-burn farming, and plowing all disrupted Neolithic environments. Some plants, animals, and insects lost their homes and food sources; niches opened for "weed species," those which thrive in disturbed landscapes and under marginal conditions. Domesticating the landscape shortened the human food chain; it reduced the number of species but increased the population densities of the survivors. Among the creatures that battened most on such change were insects, especially mosquitoes.

There are at least a few thousand species of mosquitoes, and some 10 percent of them transmit diseases to humans. As a family, mosquitoes are vigorous opportunists, but each type has adapted to specific temperatures, altitudes, and breeding conditions. While males are vegetarians, the females need blood meals to provide protein for their eggs, and they are fussy eaters. Some feed only at dusk, others only at night, dawn, or midday. Some feed only on certain animals; others make do with a variety of hosts. If humans displace or kill off the mosquitoes' usual hosts, many will gladly dine on people instead, passing on such diseases as yellow fever, malaria, dengue (breakbone fever), and several kinds of viral encephalitis.

Africa offers a frightful example of environmental disruption's inviting disease. Mosquitoes there carry the protozoon *Plasmodium ovale*, which causes a relatively mild form of malaria; it is an heirloom disease, inherited from our primate forebears. When West Africans first cleared land for farming, their slash-and-burn scarring of the shady rain forest disrupted the breeding places of the mosquito that carried *Plasmodium ovale*. Its niche was taken by *Anopheles gambiae*, an aggressive mosquito that thrives in small stagnant pools on cleared land. Its original host was birds, but it prefers human blood to any other meal. And it transmits *Plasmodium falciparum*, the cause of malignant subtertian malaria, a disease with symptoms as cruel as its name is ominous. It is responsible for at least 95 percent of all malaria deaths.

The people of West Africa survived the malarial plague their farming brought about, but only by a genetic adaptation called the sickle-cell trait. Sickle-shaped red blood cells offer *Plasmodium falciparum* a starvation diet; the result is lighter infection and fewer symptoms. Unfortunately, the trait also starves its bearer; many children born with it are prone to weakness, debility, and early death. It is a new and imperfect defense, substituting life-shortening anemia for a life-ending infection. It afflicts many people of African descent who are not exposed to malaria, including 25 percent of American blacks. Similar hereditary anemias, such as thalassemia, developed in the Neolithic Mediterranean and Near East as a defense against malaria.

This only introduces the subject of malaria, to which we shall return. It probably was not very important to humans until they created villages; then it became one of the most influential diseases in human history. Most Americans think of malaria as a tropical ill of the past, but it is neither limited to the tropics nor under control. It was long common in Europe, extending north almost to the Arctic Circle. In the United States, it nearly killed off the Jamestown colonists in 1607, and for decades after the Civil War it wrought devastation in the South and Midwest. Malaria was contained in the United States between the world wars, and in most of Europe soon after World War II, but it still kills a million children each year in Africa, and perhaps another million in the rest of the world. Its return to industrialized nations is prevented only by continuing public health vigilance.

Farming, timbering, animal domestication, irrigation, and other traumas to natural ecosystems have all created new breeding grounds for malaria's carriers. The domestication of malaria-infected birds may have helped bring additional forms of the disease to humans. The introduction of pottery vessels for storing water created perfect small breeding pools for malarial mosquitoes. Today abandoned auto and truck tires have the same effect.

Plowing, irrigation, and fertilizer invited other diseases into Neolithic settlements. Plowing began in the Old World some 5,000 years ago, irrigation a millennium later. It is debated whether swelling populations impelled advances in farming or better farming let populations expand. Whatever the case, human numbers exploded within centuries of the start of intensive agriculture in the Near East, the Nile valley, and parts of the New World. Then people began dumping human and animal wastes in their fields to save overused soil from exhaustion. The use of "night soil" made old and new diseases spread at a faster pace then ever.

The ditches and puddles of irrigated fields were ideal homes for pathogens and creatures that carried them; these caused malaria, sleeping sickness, and encephalitis. Irrigation also helped spread fecal contaminants in local waters. One of the worst new pathogens in village streams and irrigation ditches was the schistosome. This microscopic parasite entered humans from water containing infected snails; then human feces reinfected the snails. Schistosomes weaken and kill in many unpleasant ways, including bloody bladder damage and lesions of the liver and lungs. Schistosome eggs have been found in Egyptian and Chinese mummies 3,000 years old; now the disease afflicts as many as

100 million people around the world. Together, malaria, schisto-somiasis, and tuberculosis cause more sickness and death worldwide than any other three infectious diseases.

Farming, irrigation, and pollution also took a growing toll through such intestinal diseases as dysentery, shigellosis, and eventually cholera. This happened even in temperate climates. For many dangerous patho-gens, the new farming technology re-created warm, wet environments like those abandoned by early humans when they left subtropical Africa. Now the parasites had increasingly dense crowds of humans and domes-ticated animals to prey on. The result was stubborn cycles of reinfection that continued into the twentieth century. Helminth diseases and intes-tinal infections remained common in this country's rural South into the 1930s. Japan's fields were heavily fertilized with night soil into the 1950s—a practice that caused widespread sickness. In much of Africa and Southeast Asia today, people work in warm, wet fields rife with parasites. Predictably, the result is a "wormy" population, sapped of energy and disease resistance.

Neolithic farmers invited new diseases even when they stopped farming, to rest the soil or to work new areas. The abandoned fields became overgrown with scrub, an ecosystem with its own distinctive flora and fauna. Often scrub is thick with tiny microbes called rickett-siae, many of which are transmitted by ticks and mites. In the United States, the most common tick-borne rickettsial disease is lethal Rocky Mountain spotted fever. In Southeast Asia, it is scrub typhus, which afflicts many farmers there. If not combatted with antibiotics, it can kill up to 60 percent of its victims. During World War II, scrub typhus struck so many Allied soldiers in the Pacific theater that it ranked second only to malaria in causing sickness and death from disease.

Human rickettsial diseases—Rocky Mountain spotted fever, scrub-typhus, Q fever, rickettsial pox, endemic typhus, and others—probably appeared as soon as people formed permanent villages with scrubby clearings around them. They must have become more common as villagers alternated or abandoned fields and scrub species took over; ticks picked up rickettsiae from wild rodents, in which they caused only mild disease, and passed them to humans and their domesticated animals.

Some of the new diseases of the Neolithic era slowed the growth and development of human society. Population biologists believe that for thousands of years malaria, sleeping sickness, helminth infections, and diarrheal diseases limited the size and locale of human populations. In many parts of the world they are still a drag on agriculture and development. Even in many places where they are more or less controlled, they break out lethally when natural disaster, social chaos, or war disrupts modern defenses against them.

Just as Australian aborigines and Kalahari Bushmen give us an indirect glimpse of ancient hunter-gatherer life, villages in developing nations suggest how new diseases struck Neolithic settlements. A description of a village epidemic can make one think that something similar must have happened a thousand times before and gone unrecorded. Such is the case of the Mystery Disease of Pudoc, which broke out in the Philippines in 1965. The story of the disease offers a picture of a new zoonosis gripping a village society. It is the sort of improbable tale that encourages one to wander the sometimes morbid halls of medical history.

The name Mystery Disease of Pudoc does not come from a tabloid or junk film. It was coined by its victims, the Ilocano people of Pudoc West (hereafter called simply Pudoc), and adopted by researchers who came to study it. In that village in northern Luzon, people were gripped by a wasting intestinal disease that often led to a miserable death. This Ilocano misery provoked a peculiarly Ilocano explanation.

The Ilocanos were a traditional people who survived by farming and fishing. Their life probably was not very different from that of many Neolithic peoples. When a lethal epidemic struck, they believed the reason was an offense against their river god. Some villagers thought that when they had cut a mango tree near the village, a branch had fallen and killed two of the god's children. Others were convinced that a water buffalo washed up by the river and eaten by villagers had been the god's property. Whatever the reason, said the people of Pudoc, an enraged river god had condemned them to sickness and death.

This diagnosis called for a traditional cure. The village hired two of the healers and exorcists known in rural Luzon as *herbularios*. These visiting experts said that to placate the god, the people of Pudoc had to build a shrine next to the offended mango tree and pay tributes of money, food, and livestock. Under pain of death, they must not walk beneath the tree. The villagers complied, yet they kept getting sick and dying. Meanwhile, the tributes of food disappeared into the exorcists' bellies, the money into their pockets. Fear of magical retribution by the *herbularios* kept protest to a sullen grumble. After one of the *herbularios* wasted and died, the villagers found courage to throw out the other one.

Reports of deaths in Pudoc had begun trickling to Manila. Early in 1967, a year and a half after the epidemic began, a government medical team arrived in Pudoc and found a situation that might have been one of Joseph Conrad's nightmares. A third of the villagers were sick; those with advanced disease were wretched and skeletal. At least sixty people had died, a disproportionate number of them men between twenty and forty.

The medical team started to do physical exams and take histories. The disease was a chronic enteritis, or intestinal inflammation, that began with a gurgling stomach, occasional diarrhea, and mild stomachache. It progressed insidiously to severe diarrhea, sharp pain, vomiting, dehydration, weakness, severe weight loss, and often death. Exhaustive tests revealed no bacteria, viruses, or protozoa. The researchers, like the villagers, spoke of the Mystery Disease.

At last an autopsy revealed the cause, hordes of microscopic helminths ravaging the small intestine. The researchers had never seen anything like them. The new parasite was dubbed *Capillaria philippinensis,* and the Mystery Disease acquired a formal name, intestinal capillariasis. Now the questions were where did it come from, and how. Finding answers became increasingly important over the next decade, as epidemics broke out in Mindanao and Thailand, and cases appeared in Japan.

An emergency hospital and a research center were set up in Tagudin, a village south of Pudoc, and the hunt began for the parasite's natural host. Researchers tried in vain to grow *Capillaria* eggs in snails, earthworms, gerbils, and goats. Finally the eggs were found in fish in local lagoons; researchers in Thailand found that certain fish there were also susceptible. Worm larvae taken from the fish were fed to monkeys, which became infected and died. Scientists suspected that the fish were secondary hosts, not the parasite's original source. They kept looking, and in 1979 they learned that certain fish-eating birds could be infected; these were probably the primary hosts. By this time more than 2,000 cases and 100 deaths had been reported in the Philippines alone. Surely many more cases, and more deaths, had gone unreported or misdiagnosed.

How the helminth had reached humans was still a mystery. In

Tagudin and in Thailand, researchers looked at more than 150,000 specimens of local wildlife, and at thousands of samples of market food. *Capillaria philippinensis* turned up only in certain fish. The search in Tagudin focused on Ilocano lifestyle and diet, especially a dish aptly called "jumping salad"—live shrimp seasoned with vinegar, garlic, and chili peppers.

Shrimp were not the only food the Ilocanos ate so fresh that it still moved. Ilocano farmers tended their fields in the morning; by the time they reached their fish traps in the lagoons, in the afternoon, they were hungry, and they ate some of the raw fish, shrimps, crabs, snails, and squid in their traps. They liked to bite open a fish's abdomen and suck out the intestinal juices; egg-laden females were a special delicacy. Also, the men gathered at night to drink gin or *basi,* a sugar-cane wine, while they snacked on the raw vital organs of goats and cows. Predictably, they were a parasitologist's mother lode. This pattern of work and diet showed why twice as many men as women in Pudoc had caught capillariasis, especially men of working age.

It was also becoming clear why *Capillaria* infections turned up in other places. In Mindanao, as in Pudoc, villagers ate uncooked fish, snails, and crabs. The Thais who caught the disease were also rural farmers and fishermen who ate raw meat and raw freshwater fish. In all these areas, Western ideas of sanitation were a faint whisper. People and their animals defecated wherever the urge took them; during the rainy season, downpours probably washed parasite eggs from feces into the streams and lagoons. In Pudoc, people used the lagoons as fountains, laundries, swimming pools, and privies. All the affected peoples had many parasitic diseases besides capillariasis.

Two of the Pudoc researchers, John Cross and Manoon Bhaibulaya, doubt that capillariasis is truly a new disease. It may have been present in rural Philippines and Thailand sporadically for generations, ignored, undiagnosed, or misdiagnosed. Why did it suddenly become epidemic in Pudoc? Perhaps one infected person visiting from another village left parasites in the local water and thus the local fish. Perhaps migrating fish brought it to the lagoon. Perhaps migratory fish-eating birds brought it during a stop on their travels. Or the ultimate reservoir of infection may be some still unrecognized fish, bird, or mammal. Regardless, once one person has the disease, he can infect the local fish and thus his neighbors.

Cases were declining by the early 1980s; thanks to effective drugs, deaths had become rare. The disease might virtually disappear if people stopped eating raw fish, but that is unlikely in much of Asia. To health education programs, Ilocanos replied that cooking ruins the flavor of seafood; besides, they said, their ancestors lived full lives while eating raw fish every day. Privies were made available in some villages, but many Ilocanos complained of constipation from having to defecate in such unaccustomed enclosed places. At last report, some still attributed the disease to enraged gods.

Intestinal capillariasis may lurk in animal reservoirs around Asia's Pacific rim, awaiting opportunities to reappear—as perhaps it has done more than once in the past. It reminds us what village life has always offered in the way of infectious disease, and how people have unwittingly abetted the creatures that sicken them. Neolithic villagers must have survived countless mystery diseases like that of Pudoc. Probably some villages did not. Such epidemics were rude preparation for a ruder time to come, the emergence of urban life.

Four

splendor and plague

The glory and squalor of cities.
A matter of numbers. Food, famine,
and killing labor. Vectors that fly, creep,
and crawl. Vicious changelings. A plague
travels to Athens, the first of many.

From Babylon to Paris, the metropolis has evoked images of splendor and apocalypse. Thomas Wolfe, the prototypical small-town boy infatuated with big cities, called them "the places where we feel our lives will be gloriously fulfilled, our hungers fed." For variety and possibilities, they equal anything in nature. No tropical forest surpasses London or Paris in richness of life, no mountain is more stunning than Manhattan's skyline, no valley more inviting than the streets of Amsterdam. Cities promise new pleasures, risks, friends, lovers, fortunes. They are the electrical poetry of group life.

Cities, however, have also inspired a dark poetry of decay and misfortune. From Petronius to Dickens and Zola, writers have seen the metropolis as a warren of corruption, poverty, violence, and disease. The very existence of cities seems to bring fear that the hubris of creating them will be punished by disaster and collapse. They evoke images of the swamp and the jungle. The feeling has some basis in reality. From their beginnings until the twentieth century, cities have been pestholes. In fact, only when towns became big cities did massive die-offs become a regular part of human life. The author of the Book of Revelation saw the lethal side of cities: "He that is in the field shall die with the sword; and he that is in the city, famine and pestilence shall devour him."

When farmers and villagers began crowding into cities, this immunologically virgin mass offered a feast to germs lurking in domesticated animals, wastes, filth, and scavengers. Countless people were sickened and killed by previously unknown epidemics—smallpox, measles, mumps, influenza, scarlet fever, typhus, bubonic plague, syphilis, gonorrhea, and the common cold. Many of these diseases attacked with a savagery they rarely show today, demoralizing entire societies. If we are to see why new epidemics are again striking an increasingly urbanized world, we must understand why plagues and cities have always developed together.

For several million years, the main causes of human deaths were accidents and wounds. Permanent farms and villages made death by disease far more frequent. Then the population explosion of the Bronze age, 6,000 years ago, took city dwellers beyond a crucial threshold. Urban masses became sufficiently large and dense to support zymotics, or crowd diseases, what in other species are called herd diseases. For the first time, infection became humanity's chief cause of death. Despite a few respites, this would remain true in the West until this century. Infections are still the main killers in many poor nations, and they recurrently threaten the rich ones.

The reason epidemics did not take hold until urban times is simply the conditions imposed by numbers. While nomads were not free of infection, their most common diseases were chronic, not acute. Deadly epidemics remained relatively limited and infrequent for the same reasons they had been so among hunter-gatherers. People did not live densely packed together, aiding transmission of germs from one person to another. Their settlements were sufficiently far apart, and travel was sufficiently limited, to keep outbreaks of diseases localized.

Furthermore, most bacterial and viral infections with epidemic potential leave survivors temporarily or permanently immune. When such diseases did jump from an animal source to nomads or villagers, they flashed through the population; soon most of the people in a community were either dead or immune. The microbes, having run out of susceptible hosts, died off. Only years or generations later could the germ attack successfully again, depending on a new crop of susceptibles and another accident of reintroduction.

Writers often speak of crowd diseases with the metaphor of fire, and of human hosts as fuel. If there is too little fuel or if it is too thinly scattered, the blaze sputters out. The image, though simplistic, is basically accurate. In epidemiologists' terms, a zymotic persists only if the population is dense enough to keep transmitting the germs and big enough to keep producing new susceptibles. Herd diseases jump from animals to humans and thrive only if the people form a superherd. Once cities hold several thousand people, they can support most present-day crowd diseases. That first happened in the ancient Middle East.

The census is a rather new invention, and an imperfect one. No one knows precisely the population of the United States or the world today, and the further back one looks in time, the fuzzier guesses become. We do not know which cities of Mesopotamia were the first big enough to sustain the wildfire of various zymotics, each of which has its own population threshold. Perhaps it was Ur or Nineveh; perhaps it was Babylon, whose name became a symbol of city life's pleasures and perils. Still, the trends of ancient population growth are clear, the major milestones visible.

In the early Neolithic, cultivated crops and herds allowed a surge in village populations; that in turn demanded still more productive farming. Once established, the cycle of growing populations and bigger food supplies spun faster. By one estimate, the world's population grew from 4 million 12,000 years ago to 5 million 5,000 years later. By the late Neolithic, irrigated fertile plains around the world supported cities as large as 100,000 people. These first appeared in the valleys of the Tigris and Euphrates in Mesopotamia, then along the Nile in Egypt, the Indus in India, and the Yellow (Huang) in China. Intensive farming and big cities would arise later in Peru and in Mesoamerica.

Modern readers and museumgoers are fascinated by these cities' monuments and art. It is easy to ignore the mundane fact that the gardens of Babylon, the golden mask of King Tutankhamen, and the masonry walls of Machu Picchu all rested on one essential asset. Big cities and their culture could not exist without mountains of food—that is, without farmers who produced many times what they needed to sustain themselves. After people added tin to copper and made tools of bronze, food production could really soar, and urban life could boom.

The Bronze Age began in Eurasia 6,000 years ago, the Iron Age 3,000 years later. Improved metal tools changed farming, warfare, construction, and more; the terms "Bronze Age" and "Iron Age" denote not only the use of those metals but the technologies and social organization that developed with them. Metal plows furrowed the land; metal scythes cut crops, and metal flails threshed them; metal axes felled forests and built cities and ships. Food production and population growth cycled faster and faster. By 2,500 years ago, the world's population had risen from the Neolithic 5 million to 100 million. In the next seven centuries, it doubled.

The first big-city residents lived amid splendor and abundance their foraging ancestors could not have imagined. Their fields and herds supported walled cities with temples, gardens, and intricate water systems. In good years, their granaries bulged with huge stores of barley and wheat. Their cattle and sheep provided meat, dairy products, and hides; oxen pulled plows and turned millstones, horses drew chariots. Houses contained woven cloth and pottery vessels. The people had written languages, recorded histories, codes of law, and of course tax rolls. Compared with hunter-gatherers, says historian Alfred Crosby, they were billionaires.

But as billionaires go, the majority were neither healthy nor long-lived. As happened early in the Agricultural Revolution, health tended to retreat while technology advanced. There were some exceptions; skeletons from Bronze Age Greece and Iron Age Scandinavia show that many people were as tall and robust as hunter-gatherers. But in much of Eurasia and the New World, urban nutrition and life span continued the decline that had begun with village life.

One reason was that wheat, rice, and maize made up half or more of many people's diets. This dependence on starchy staples put urbanites at the mercy of crop blights and bad weather. Unlike nomads or villagers, they could not move on when fields were ravaged by drought or soil exhaustion, crops stricken by fungus, herds felled by epidemics, stored grain ruined by molds. There must have been terrible famines, with deaths from such deficiency diseases as scurvy, rickets, pellagra, beri-beri, and kwashiorkor. These would have caused death not only directly but indirectly, by inviting and amplifying infections.

To make bad things worse, the toll of malnutrition and disease was spread around less evenly than before. As social complexity grew, classes and castes arose; so did the gulf between rulers and slaves, priests and laborers. On this point ancient bones are eloquent; the upper classes had more varied diets and consumed more protein than the poor. Even in relatively healthy Bronze Age Greece, royalty were better nourished and longer-lived than commoners. In the great societies of ancient Meso-america, few besides nobles and priests were spared frequent malnutri-tion and sickness.

Another source of illness was the trend toward specialized labor. Many trades and crafts carried distinctive health risks. In the past, individual hunters may have been exposed occasionally to anthrax by killing and skinning wild sheep. Now, in big cities, many workers were exposed to anthrax all day, every day. The anthrax bacillus can survive for years despite heat, cold, or dryness, in infected wool and hides, even in contaminated earth and dust. This favorite weapon of today's germ warfare engineers was a daily risk to Bronze Age herdsmen, butchers, shearers, wool handlers, and tanners.

Anthrax was only one of many zoonoses to gain a foothold among city workers. For millennia, many people in the Middle East used dog feces in tanning hides; that made tanners and shoemakers prey to echinococcosis, a helminthic disease often passed from dogs to humans. Bronze Age workers probably also caught glanders, an infection of horses that caused deadly pneumonia in humans. Food handlers were imperiled by many fevers, since stored food inevitably holds the germ-laden wastes of rodents and insects.

Just as virtually every step toward agriculture had invited new diseases, so did each new technology and condition of urban life. Doctors and burial workers were obviously at high risk for disease. So were people who labored in the water systems of Iron Age cities. Potters and miners, poisoned by working with such toxic metals as mercury, lead, and arsenic, must have been highly susceptible to infection. And everyone was at new risk because crowding, filth, and pollution were on the rise.

The gardens of Babylon and temples of Egypt were emblems of urban glory, but the alleys in their shadows were choked with garbage. Homes reeked with fetid air and smoke. Vast amounts of human and animal wastes accumulated; water was drawn from contaminated wells, food harvested from tainted fields. Dirt and refuse drew every germ-bearing scavenger that flew, crept, or crawled. Puddles, cisterns, and water vessels harbored mosquitoes carrying malaria, yellow fever, dengue, and encephalitis. Granaries and sewers stirred with rats that carried typhus, relapsing fever, hemorrhagic fevers, and perhaps bubonic plague. Householders caught toxoplasmosis parasites from pet cats. Leptospirosis infection from dogs gave humans a deadly form of meningitis.

This explosion of crowd diseases and other new infections shows why scientists often categorize germs not by the diseases they cause but by how they are transmitted. In his classic book on plagues, William McNeill says that if there is any conceivable way a germ can travel from one species to another, some microbe will find it. That happened constantly in the first big cities. Each of the four major types of disease transmission—airborne, waterborne, direct contact, and by insects or other vectors—was enhanced by urban life.

About half of all human infections are spread by microscopic droplets breathed, coughed, or sneezed into the air. Crowded, poorly ventilated ancient cities were Eden for such diseases, from measles and mumps to tuberculosis. The most common of all crowd diseases, the common cold—more accurately colds, similar conditions caused by a hundred or so related viruses—is airborne. Since they infect only humans and horses, these viruses doubtless are descended from some grandmother of all colds that made the leap from domesticated horses to people four or five thousand years ago.

Disease germs that spread by direct body contact became more common and severe. Skin diseases are common but usually mild among villagers in warm, moist climates; they are transmitted easily among playing children by the touch of perspiring skin. As settled populations grew, especially in temperate climates, full-body clothing became customary year round; germs that infected the skin found their usual homes and means of transmission threatened. They took refuge in warm, moist parts of the body, where they could survive during longer intervals between skin contacts of hosts. That is, they settled in and around the mouth, genitals, and anus, awaiting sexual transmission. Thus they became diseases not of children but of adolescents and adults. In cities, the rise in each person's number of potential sex partners allowed gonorrhea and syphilis to become rampant. Prostitution, a regular trade at least several thousand years ago, created a new pool of people who were at once victims and sources of sexually transmitted diseases.

Waterborne germs are transmitted by drinking and bathing and by tainted food, fingers, and household implements. Their usual ultimate source is human or animal feces, which Bronze Age cities produced in prodigious amounts. Fecally contaminated water can spread polio, cholera, viral hepatitis, whooping cough, diphtheria, typhoid, and paratyphoid fever. Most of these diseases adapted to urban people from their original homes in domestic animals and scavengers. Typhoid, for instance, is often fatal to humans, but rarely severe in rodents and birds, which are thought to be the original hosts.

Typhoid is one of many diseases caused by *Salmonella* bacteria; literally thousands of types of these bacteria live in warm-blooded species. Well into this century, *Salmonella* infections were a leading cause of fatal infant diarrhea. They still cause food poisoning in people of all ages, mostly through poultry and eggs, but sometimes through vegetables washed in polluted water or touched by unwashed hands. Similarly, contaminated milk, meat, and vegetables can transmit diphtheria, brucellosis, tuberculosis, and other infections.

Another nasty waterborne germ is the *Shigella* bacterium. It has no known animal reservoir but primates; it probably adapted to man from monkeys that raided farmers' crops or that lived near people, as the temple monkeys of India still do. Shigellosis remains common in Africa, and it causes more intestinal infections in developed nations, including the United States, than was once supposed. It has also become a sexually transmitted disease, thanks to a variety of anal-erotic flippancies—a warning that a zoonosis, having found one means of human-to-human transmission, may find still others.

Finally, infections are spread by vectors, creatures that carry them to other species, with or without being affected. Rodents, birds, and snails can carry pathogens to humans, but by far the most varied and

numerous vectors are arthropods—organisms with jointed external skeletons such as insects and ticks. Early cities and their disrupted environs
gave new homes to countless arthropods, such as beetles, bedbugs,
mites, and mosquitoes. They invaded fields, gardens, houses, granaries,
refuse heaps, animal pens, even clothes, bedding, and dust. Among the
arbo (short for "arthropod-borne") infections that adapted to humans
are malignant tertian malaria, yellow fever, dengue, sleeping sickness,
typhus, typhus's rickettsial relatives, and the recently arrived Lyme
disease.

Some of those germs adapted exclusively to people; others remained
able to infect their original hosts and vectors as well. Being at home in
two or more species gives a germ obvious advantages. It is more likely to
survive the temporary or prolonged absence of one type of host; and a
mobile host or vector, such as a bird or mosquito, can extend the germ's
geographic range and its number of victims. Such adaptability is a
double lease on life in times of ecological disruption. And the growth of
early cities and their surrounding fields created disruptions on a monu-
mental scale.

When Bronze Age cities were inviting so many new germs and
means of transmissions, some social customs may have retarded the
spread of infections slightly. The Hebrews expelled individuals with
severe skin diseases and, like Hindus and Muslims, required frequent
ritual washing with water or sand. Historians have guessed that some
dietary laws, such as the Hebrew ban on pork and shellfish, rose from
knowledge that these foods could pass on diseases such as trichinosis
and hepatitis. It seems more likely that these practices reflected ideas of
ritual purity or magical taboo; the idea of infection in its modern sense
was unknown.

Even if Bronze Age people had known why new diseases were
striking them, that knowledge probably wouldn't have helped much.
They still would have had no cures and few preventives. Population
pressure still would have forced them to disrupt their ecosystems by land
clearance for farming and herding, and to change their daily lives with
new technology, trade, and living conditions. Anyway, they did not know
where new diseases came from; they could explain them only as divine
punishment, and pray in vain for relief.

From about 4000 B.C. to A.D. 400, the onslaught of new diseases
continued, as people were exposed to germs they had met rarely during

the millions of years when their immune systems had evolved. Fortunately, these did not all appear at once, in part because each has its own population threshold. But once those thresholds were breached, epidemics swept through cities and then spilled into surrounding towns and villages. The villagers had been exposed to fewer germs as they grew up, so their immune defenses were, in the metaphor of epidemiologists, naive. Therefore they may have been hit even harder than city dwellers.

The history of urban epidemics begins with fragmentary accounts in the ancient writings of the Sumerians, Babylonians, Hebrews, Hittites, Egyptians, Greeks, Romans, Indians, and Chinese. Some of the oldest and most familiar mentions in the West are in the Old Testament, which has several episodes of what sound like devastating new diseases of humans and animals. Around 1500 B.C., we are told in Exodus, God cast a plague on the Egyptians, who held the Hebrews in bondage; the plague brought "sores that break into pustules on man and beast." In Deuteronomy, God assures the Hebrews that when they return to their homeland, He will "take away all the sickness from you . . . the foul diseases of Egypt which you know so well." The promise may have come true for a people leaving the humid, crowded cities of the Nile for a drier, more sparsely settled country.

A puzzling and bizarre story in the First Book of Samuel relates that the Philistines, having seized the Hebrews' holy ark, were laid low at Ashdod by pestilence. They concluded that instead of continuing to die by the thousands, they had better return the ark, and perhaps the pestilence with it. They did so, and the Hebrews began to die of the plague, allegedly for having presumed to gaze upon the ark. The epidemic killed more than 50,000 of them. Its identity has been the subject of guesses including smallpox and bubonic plague. Some scholars favor the latter because of the peculiar sacrifice five Philistine lords made when they returned the ark, five golden mice and five golden emerods.

Now, the word "emerods" is usually said to mean hemorrhoids. Some historians claim it also meant buboes, the swollen lymph glands typical of bubonic plague. The mind boggles at trying to visualize five golden buboes, let alone five golden hemorrhoids. Is one to picture them carried in procession on a platter, on velvet cushions, or in the raised hands of bearded and lamenting lords? Believers in the bubonic theory point to the golden rodents as evidence, but it is rats, not mice, that carry the plague. In any case, both the Hebrews and the Philistines lacked the concepts of infection and contagion, so they probably would not have linked a disease with the carrier of its germs. So the bubonic theory is no better than others.

Intellectual hobbyists have created a large literature of such guesses, trying to pin tails on ancient epidemics. Some of these writings are brilliant, some competent, and many merely obsessive or crankish. Every major disease, and many a minor one, has a following, a sort of scholarly fan club, devoted to revealing its hidden history and proving its ubiquity and importance. Unfortunately, aficionados of malaria, bubonic plague, rheumatoid arthritis, or influenza tend to see their favorite affliction everywhere in history. On slender evidence or none, they list monarchs and empires it has crippled or buried. Other writers may even quote their speculations as facts.

Diagnosing the diseases of antiquity is much harder than diagnosing live patients, and that says a lot. The margin of diagnostic error in modern teaching hospitals ranges from 10 to 20 percent and even higher. We can only speculate how much greater it is in biblical and classical scholarship. Part of the problem is the nature of the evidence. Today we classify and describe diseases according to anatomy or cause—as, say, diseases of the kidneys or infections caused by streptococci. Sometimes ancient descriptions of symptoms match modern ones; the Hippocratic texts of ancient Greece portray malaria and tuberculosis in ways any doctor would recognize today. But many writers of ancient Europe, Egypt, and Asia described diseases according to the gods, miasmas, or humors they thought caused them. One can labor long over the Egyptian Ebers papyrus, the Latin of Galen, or the Arabic of Rhazes and be lucky to link their categories with modern disease concepts. And since many diseases have evolved and changed over the millennia, one is doubly lucky to connect an ancient plague with a modern one.

When zymotics first raged through ancient cities, they often took forms we would not recognize. A new zoonosis is usually a vicious changeling. As a germ attacks a new species or a virgin population, it tends to strike people of all ages, damaging many organs and causing florid symptoms. As host and pathogen adapt to each other, over centuries or sometimes even a few generations, the disease tends to become less acute, with milder symptoms in fewer organs, and eventually subside to a common childhood disease. Even its means of transmission may change.

Syphilis and measles are good examples. When syphilis first appeared in Europe, in the late fifteenth century, it caused repulsive pustules over the entire body and attacked many internal organs; often it brought death within a few years. Just a half-century later, symptoms were usually limited to the genitals, face, and nervous system, and a patient might live for many years, even decades. Similarly, the early epidemics of smallpox, mumps, measles, and scarlet fever must have been far more virulent than the diseases we know by those names. The standard example of such a changing zymotic is measles.

The measles virus probably evolved from the one that causes distemper in dogs. (They had inherited it from their ancestors, the wolves.) Since dogs are man's oldest animal companions, individuals may have caught some form of the distemper virus before cities arose, but the disease could not have lasted long or recurred regularly. Transmitted by coughing and sneezing, it would have run quickly through groups large or small; although the death toll might have been high, survivors would have been immune to reinfection. Epidemics were doubtless infrequent and self-limiting.

Eventually the virus adapted to humans, in a variant that could no longer survive in dogs. This new human disease, with no animal reservoir, needed a constant supply of new human susceptibles to keep from burning out—probably at least 7,000 at any given time. These susceptibles could result only from new births or from migrations into cities. In recent centuries, measles has died out on islands with fewer than 500,000 people unless reintroduced from the outside. In an area with a high birth rate and much in-migration, the smallest population that could have sustained measles in ancient cities was probably a few hundred thousand—the highest population threshold of any common zymotic.

Therefore measles could not have become part of human life until three or four thousand years ago, in the late Bronze or early Iron Age. In the West it eventually became a routine childhood disease and was almost eradicated by vaccination in the 1970s. Recently it has reappeared because of failure to immunize all children, especially in inner-city slums. In some non-Western nations, measles is still widespread, and very damaging. Poverty, malnutrition, crowding, poor hygiene, and lack of medical care make the disease more common and more severe than it would otherwise be. Even in a prosperous nation, measles can leave children dead, brain-damaged, or suffering secondary infections such as pneumonia. In some poor countries, the death rate from measles is still above 10 percent. Measles still kills a million people a year, most of them children in the Third World.

When measles first attacked big-city dwellers, it was doubtless far worse than it is in the West today, but evidence is sparse. Ancient medical writings usually lumped together all exanthemata, or infections that cause rashes, such as measles, smallpox, chickenpox, scarlet fever, rubella, and erysipelas. The Persian physician Rhazes distinguished measles from smallpox in the tenth century, but he thought they were two phases of one disease. Indeed, epidemics of measles and smallpox, both viral crowd infections without animal reservoirs, have often traveled on each other's heels. Only in the sixteenth through eighteenth centuries were rash diseases all clearly distinguished. Among the last was rubella, described by a German doctor and therefore known as German measles.

Hippocrates did not mention anything resembling measles, but measles and smallpox probably existed in his day in other parts of Eurasia. By the early Christian era, they probably were entrenched all over Europe. William McNeill guesses that the terrible Roman plagues of A.D. 165–180 and 251–266 were the Mediterranean world's first virulent attacks of those diseases. The Greek physician Galen, who spent much of his life in Rome, witnessed that first plague, which he said caused high fever, a rash, and spitting of blood. At the height of the second plague, as many as 5,000 people died in Rome each day. Two centuries later, terrible rash-producing epidemics raked the Roman and Chinese empires, killing so many people and bringing so much social disruption that they must have contributed to both empires' downfall.

By the sixteenth century, European familiarity with measles and smallpox had made those infections common in childhood, with relatively low death rates. The conquistadors carried them across the Atlantic, and for 200 years they killed millions of Native Americans of all ages. The devastation of measles alone can be guessed from its more recent attacks on virgin populations. When it first hit Fiji, in the 1870s, it killed 20 to 30 percent of the native people. It had similar virulence when it was introduced in Samoa, Hawaii, and West Africa.

All the acute rash diseases except chicken pox are products of big cities and their human superherds. As we shall see, measles and small-pox have special historical importance, because they acted as biological weapons of colonialism. Another rash disease, the great plague of Athens, has fascinated historians because it helped destroy the city's Periclean golden age. It is equally important as a sign that a new and perilous stage had arrived in human health and illness.

In 430 B.C., when Athenian culture and power were at their height, the city had drifted into war with Sparta and its Peloponnesian allies. The walls of Athens and its port, Piraeus, held more than 200,000 inhabitants. The Spartan invasion sent uncounted villagers pouring in for refuge; they were crowded into hot, stifling huts. Soon the city was scourged by a plague unlike any Athenians had ever seen, with cruel symptoms and stunning mortality. The war's great chronicler, Thucydides, who survived the disease, left a harrowing account in the hope that posterity would recognize the killer if it ever attacked again.

The plague, he said, began in Ethiopia and spread to Egypt and Libya. It must have arrived in Greece by ship from Egypt; it appeared suddenly in Piraeus and spread to besieged, overcrowded Athens, spar-ing most of the Spartans outside the city walls. People became feverish and flushed, sneezed, and developed a "bloody" tongue and throat. Then came coughing, vomiting, and diarrhea that prostrated the vic-tims. Their bodies broke out in sores that became ulcers; sleepless and agitated, unable to bear the touch of clothes or bedding, they staggered naked through the streets, seeking water for their unquenchable thirst. They died in streets, in temples, in wells into which they had fallen. Corpses heaped up, along with the bodies of dogs that had died with their masters. Most birds and carrion eaters shunned the bodies; those that ate them died. The disease was a lethal stranger even to Athens' pets and scavengers.

The plague killed young and old, rich and poor, slaves and generals and physicians. It subsided briefly, only to race again through the city and attack the fleet Athens had sent against Sparta. Pericles, who was with the fleet, died, as did more than a quarter of its 4,000 soldiers. The worst of the plague lasted two years, and it lingered for almost three more. By then it had killed one-third of Athens' people. Many survivors had lost fingers, toes, eyesight, or memory.

The sufferings of the plague, Thucydides wrote, were beyond human endurance; society came unglued. License, lawlessness, and even murder went unchecked. The poor seized the fortunes of the dead and tried to squander everything at once; no one expected to live to reach trial. Besides, people already felt sentenced to a punishment worse than any court's. The plague unraveled not just Athens' military strength but its civic and moral fabric.

With its great naval power, Athens should have been able to outlast Sparta. Without the plague, it might have. It fell only after bleeding through five years of plague and almost thirty of intermittent war. Athens would never fully regain its political or cultural glory. More than 2,000 years later, the West's finest minds would still dream of re-creating the golden age that the plague had helped destroy.

The symptoms of that plague match no illness known today. It may have been the violent European debut of measles, scarlet fever, small-pox, typhoid, or some disease that no longer exists. Another guess is that an epidemic of influenza or something like it was complicated by staph-ylococcus infections, producing a toxic shock syndrome. We will never be sure.

It is chilling to think that such a plague was probably no worse than many other epidemics of the Bronze and Iron ages. But more important for history than its virulence was its origin. According to Thucydides, it came by ship from Africa. If so, this is the first recorded instance of the spread of a crowd disease from one of the world's major centers of infection to another.

Long before the golden age of Athens, foci of infections had devel-oped in Mediterranean Europe, Egypt, Mesopotamia, India, and China. Each area, with its own climate, ecosystem, and germs, had a distinctive set of infections, to which people adapted with a distinctive complex of immune defenses. By the time of the plague of Athens, the foci of Egypt and Mesopotamia had probably fused and come into contact with that of India. India's may have already brushed China's. But only with Thucydides' account do we have a documented episode in the emer-gence of a band of common infections across Eurasia and North Africa. For pathogens, these foci were merging and becoming one. The reasons were war and travel.

There are other ancient mentions of plagues, and those records of devastation and despair hold signs that a common pool of zymotics was emerging across the urbanized Old World. In the fourteenth century B.C., a Hittite priest bemoaned a plague that had been killing his people since their contact with Egyptian prisoners of war twenty years earlier. He wrote that "men have been dying in my father's days, in my brother's days, and in mine. . . . The agony in my heart and the anguish in my soul I cannot endure any more." Fields went unplowed and bread unbaked, because farmers and grinding women were dead. Sheepfolds and cow pens were empty; the shepherds and cowherds had perished. The priest prayed in vain to a river god he thought had sent the plague to punish the Hittites for failing to keep a wartime pledge to the deity. His lament, still poignant across 3,400 years, tells us that the Hittite plague, like the plague of Athens, was the poisoned fruit of military power.

More and more epidemics were spreading across the world by travel, trade, and war. There were more people, bigger armies, better transport. By 6,000 years ago, the Egyptians had boats that could carry cargo and troops; 500 years later, so did the Mesopotamians. Knossos, in Bronze Age Crete, was the hub of a trade network connecting Europe and the Near East. By Roman times, cargo ships weighing more than a thousand tons traded between Italy and India. People carried diseases by ship and camel caravan, across seas and mountains and deserts; with them went stowaway pests, the germs of ship rats, and livestock parasites. By late Roman days, every major urban society in the Old World probably had passed zymotics to several others, causing the sort of devastation that had struck the Hittites and Athenians.

Diseases kept spreading because humans responded to their population explosion just as other social species do. They fought for more resources, sought new territories, tested neighbors' boundaries. Some migrated westward from India across Europe, carrying the Indo-European languages. In the Middle and Near East, the clash of growing populations brought the rise and fall of Sumerians, Akkadians, Babylonians, Kassites, Hittites, Egyptians, Assyrians, Chaldeans, and Persians. Later, tribes that would be called Germanic and central European pushed at the boundaries of the Roman empire and finally broke through. Everywhere, venturesome hordes and armies visited new diseases on lands they occupied, and tribal masses fell victim to infections of the cities they conquered. To microbes, both sides in these struggles were just fresh meat.

As fearsome as new epidemics were in the Bronze and Iron ages, less dramatic endemics probably caused just as much damage. Malaria and schistosomiasis flourished in the farms and villages that now covered more and more of the earth to feed cities' appetites. This was especially true where warm climates combined with irrigation to create something like the ideal environment of many parasites, that of the rain forest.

The signs of these ancient infections are ubiquitous. The mud-brick walls of ancient Mesopotamian cities contain shells of the *Bulinus* snail, which harbors the schistosome. Greece and Rome suffered badly from malaria; Hippocrates and Galen described three of its four varieties with perfect clinical clarity. Then, as today, endemic malaria and schistosomiasis created a debilitated peasantry, many of them too sick to work, victimized by other epidemics, and destined to early death. By draining people's vigor, such endemics stole the cities' lifeblood.

As cities enhanced old diseases and fostered new ones, births could not keep up with deaths. To replace the casualties of crowd diseases and maintain their populations, cities needed a constant inflow of migrants from the countryside. There were always willing farmers and villagers; when diseases took their rural toll, many of the men who survived fled hopefully to the economic promise of cities. One sees a result of this today in many Third World countries, where only the old, the sick, women, and children remain to work the land. Many ancient empires must have fallen because the peasants were sickly, their ranks thinned by migration to cities harrowed by epidemics.

It is a testimony to human vigor and adaptability that city life flourished despite plagues, famines, wars, and migrations. The assault by new crowd diseases was fierce and unrelenting. A human population usually needs at least a century or so to stabilize in response to an unfamiliar infection. Through the Bronze and Iron ages, people had to endure that struggle again and again, often with only brief respites. No sooner did they reach relative tolerance of a new epidemic than another arrived, caught from an animal source or a foreign focus of infection.

Cities not only survived, they grew in size and number. Catastrophic epidemics were followed by resurging populations and renewed urban growth. There are economic and biological explanations, based on models of population dynamics and resource exploitation; some of these are plausible and probably partly true. I suggest, however, that the primary force driving urban growth was something in the human temperament—a combination of curiosity, inventiveness, and hunger for stimulation.

Laboratory rats and monkeys will exhaust or even kill themselves if they are allowed to keep artificially stimulating themselves, whether with sweets, energizing drugs, or electrical jolts to the brain. We, too, are fools for stimulation. The variety and excitement of city life are super-stimuli. Humans, with much greater capacities for excitement and boredom than other creatures, have a hard time resisting those stimuli. Like rats or rhesus monkeys that will binge on sugar, cocaine, or brain shocks despite the noxious results, people risk discomfort or even death to be in environments full of challenge and adventure. Evolution has encouraged us to do so by sharpening and prolonging the curiosity and playfulness that in most species end with childhood.

But the human immune system is not infinitely adaptable. The biological price of urbanization, trade, travel, and war was increased saturation with infections across much of the Old World. Technology had allowed diseases to multiply and travel, but it could not yet offer ways to combat them. Finally the human burden of disease became so great that only the end of urban life could relieve it. Humanity experienced a population crash, of a kind some scientists fear may once again lie in our future.

ruthless cure

Wallowing toward ruin. The unblinking
Malthus: When births exceed resources.
Climax and crisis. The first pandemics.
East blights West. The coup de grâce;
follow the flea. Cured of urban
progress. A dark reprieve.

In the late Roman era, nature produced a cruel cure for the new diseases of urban life. All over Europe, Asia, and North Africa, too many people were too densely crowded; the creation of farms, roads, and cities disrupted the environment; diseases traveled with lethal effect among the once distinct disease pools of Europe, Africa, India, and China. New technology had allowed populations to explode, but it offered no cures for the resulting epidemics. Eventually spasms of pestilence brought a worldwide population crash that seemed to threaten humans with extinction. To some modern biologists, such mass die-offs seem to be natural events, but until rather recently, the usual explanation for the first pandemics was not biological, it was moral.

The moral theme appeared early in the age of pandemics; it was encouraged by the fact that plagues paid dividends to Christianity. The new faith, with its contempt for comfort and for life in this world, and its hope of resurrection in a better one, profited from infections that scourged the Roman world from the second century A.D. on. In 252, a devastating epidemic struck Carthage, in Roman North Africa, sowing death, panic, and chaos. St. Cyprian, the bishop of Carthage, wrote that each day hundreds of terrified people flocked to him for baptism. Although Christianity could not save the body, it gave hope for the soul and made sense of catastrophe.

Proselytizers preached that sin, selfishness, and paganism were the reasons people saw their families and neighbors dying each day. This appealed to a deeply rooted human tendency to see illness as punishment, plague as divine vengeance. From the ancient Hittites and Hebrews to the villagers of Pudoc and Americans with sexually transmitted diseases, epidemics have roused fits of guilt and, sometimes literally, of self-flagellation.

The early Christian claim persisted that Rome had wallowed its way to ruin. When Edward Gibbon wrote his *History of the Decline and Fall of the Roman Empire* in the late eighteenth century, he polished the old image of pagan decadence luring humanity down the slope to social and physical destruction. It remains inherited wisdom today. This alleged lesson of Rome to posterity has been applied to many times and places: sex, selfishness, and prosperity itself are agents of spiritual and bodily degeneration, and finally of death. In its crudest form, the attitude survives in Hollywood visions of men in togas and women in nothing, indulging in feast and orgy as Rome burns, volcanoes erupt, and barbarians beat at the gates. Not long ago, such scenes were conjured up as warnings against contraception, central heating, social welfare, the forty-hour work week, and other novel niceties. Rome's fall, though, was not caused by surfeits of food and sensuality but by nature's response to overpopulation and ecological change.

The ecological view of Rome's decline had its origin in Gibbon's near contemporary Thomas Malthus. In 1798, this cleric and economist produced *An Essay on the Principle of Population*, the founding classic of demography. It was a book of vast and lasting influence. Malthus argued that unchecked population growth tends to outstrip gains in food and other vital resources. When this happens, the birth rate falls and the death rate rises. The deaths result from overcrowding, poverty, famine, war, and finally disease, as "sickly seasons, epidemics, pestilence, and plague advance in terrific array, and sweep off their thousands and tens of thousands." Millions would have been more accurate. Malthus may have been right, though, in asserting that the riders of the Apocalypse were not supernal scourges but the redress of a natural imbalance.

It was not a consoling view, but Malthus was no optimist. He doubted that humanity would ever escape cycles of overpopulation and disaster. Still, he urged trying, by limiting births through sexual abstinence and late marriage. He set an example by remaining a virgin until he wedded, at age forty. He had already fathered, if nothing else, a movement, the Malthusians, who propagated his ideas; from them sprang neo-Malthusians, who called for population control but did not limit sex to reproduction; and from them came the modern birth control movement and such groups as Zero Population Growth.

Malthus was not alone in seeing links among overpopulation, epidemics, and deaths. During the nineteenth century, naturalists noticed regular cycles of increase and die-off in birds, rodents, and ungulates. Eventually they saw that population changes in prey and predators, and in parasites and hosts, were interdependent. In this century, after evolutionary theory had suggested how such patterns developed, the science of ecology emerged. The terms "ecology" and "environment" have become emotional and political slogans, but to scientists ecology is neither a mood nor a movement. It is the fascinating and sometimes sobering study of how plant and animal species in an environment do a continual dance of mutual adaptation.

There is an axiom of ecology that all the species in an environment strive for a balance, the climax state, in which they might all survive indefinitely. Each creature has a stake in that equilibrium. If a deer population explodes, for instance, the animals will denude a forest, destroying not only their own food and shelter but those of other species. If a virus infects a flowering plant, problems may rise all the way up the food chain, to bees that feed on flowers, birds that eat bees, small mammals that eat birds, and big predators that eat little ones. Knowledge of this interdependence, not sentimentality, makes scientists fidget about endangering even one apparently trivial species, for it is a thread in a fabric. A climax state is the product of long evolution; it is complex and vulnerable to disruption. A dry winter, a cool spring, the appearance of one fungus, insect, or scavenger can wreak instabilities with unknowable consequences. For viruses, humans, and every species in between, the climax state easily gives way to crisis.

The route from population crisis back to balance can be piteous. In rats, overpopulation and crowding create a biological slum. Social and aggressive behavior become abnormal; adults fight among themselves and neglect or attack the young. The birth rate falls, and infant mortality rises. Constant stress, as Dr. Hans Selye discovered, damages the circulatory system, kidneys, and immune system. Some rats die directly from stress and its behavioral effects; others succumb to it indirectly, through infections. Something similar seems to happen to other mammals, including humans. Under stress we probably are more vulnerable to many disorders, such as heart attacks, certain viral infections, and perhaps some forms of cancer.

Nature often seems to correct a severe population crisis with a mass die-off. The most efficient way to do so is by means of an infectious disease; frequently the height of a crisis is marked by epidemics that kill half or more of the population. Naturalists have seen tuberculosis kill off voles in Wales, rinderpest wipe out wild ungulates in Africa, and other epidemics reverse population explosions of ducks, squirrels, mice, muskrats, and cattle. Individuals that fight off or survive infection form the core of a smaller population with better resistance to the disease. The disease may remain potent enough to act as a gentler population control, becoming endemic in later generations and culling some of the less fit.

Human population crises and die-offs have happened from the Neolithic to the present, from England to China to Mexico. People multiplied beyond their resources and suffered famine, migration, war, and pestilence. Epidemic disease was usually the crowning disaster; for some 10,000 years, it has been the biggest cause of deaths. However, there was something distinctive about the epidemics of 1,500 to 2,000 years ago, when the entire Roman empire was ravaged by crowd diseases. It was the first time a die-off struck much of the world at one time. The first era of pandemics set off spasms of terror, and with good reason.

It is one thing to theorize about ecological balance, another to live through a plague. People feel that they are suffering a supernatural blow and await the end of society, even the end of the world. With passionate contrition they wonder what they have done to deserve it, as individuals and as a people. Such terror and guilt are as old as epidemics, but they were heightened in the first age of pandemics, as people noticed the shorter intervals between plagues that spanned the known world.

Countless local crises and die-offs are lost in prehistory. One may have happened in the second millennium B.C., when natural disasters, war, and perhaps pestilence wiped out the Bronze Age culture of Mycenaean Greece, ushering in a long "dark age" of reruralization. The Old Testament suggests that around that time, plagues also sheared back urban populations in the Near and Middle East. Similar crashes may have happened later in central Europe, from the Danube valley to the Rhine, after the shift from village life to fortified cities.

As the Christian era began, populations were hitting new highs all over the urbanized world. By conservative estimate, the world total had risen from 5 million in 5000 B.C. to 100 million in 500 B.C., and nearly twice that in the second century A.D. By then the city of Rome had more than 2 million people. Its empire, which touched Scotland, the Sahara, and the Persian Gulf, held more than 50 million.

The empire's big cities were not hopelessly unhealthful. Rome and some provincial capitals had extensive water and sewage systems, flush toilets, public baths, and food inspection. Their size and wealth made ancient Babylon and Nineveh seem paltry, and their better-off citizens lived a decade longer than had their Paleolithic forebears. Yet all the triumphs of food production, engineering, and sanitation could not prevent epidemics that made the population level off and start to decline. For several centuries the empire was undermined by conflict and chaos, and depopulated by waves of new diseases. Some came from other species, as they always had, but some came from other empires.

Before Roman times, infections from distant ecosystems had struck the West occasionally; the first recorded instance was the plague of Athens, in 430 B.C. Only thirty-five years later, the Roman world suffered a similar epidemic. It, too, came from North Africa by ship during a war, when Carthaginians besieged the Sicilian city of Syracuse. The epidemic took so many lives that almost five centuries later, Livy described it as one of the eleven major plagues that Rome had suffered since its founding.

In A.D. 79, another new disease struck Rome, possibly anthrax but probably the acute *falciparum*-borne form of malaria, complicated by some other contagion. This disease, native to Africa's rain forests, had traveled down the Nile to the Mediterranean, then spread east to the Mesopotamian Fertile Crescent and north to Greece. Greek traders and colonists eventually carried it to Italy. Roman soldiers and merchants would help several forms of malaria move as far north as England and Denmark.

For 2,000 years, wherever Europe had human crowds, new forts and settlements, domesticated animals, and standing water where mosquitoes could breed, malaria would flourish. It left people weak, apathetic, and short-lived. It may have contributed as much to Rome's fall as any epidemic that followed it. The malaria epidemic of the year 79 so devastated the Campagna, the fertile marshy area where Rome's crops grew, that its fields lay fallow and its villages stood empty. The Campagna would remain sparsely settled for much of the next two millennia. Malaria was eradicated there only in the late 1930s.

From the second century on, new diseases appeared in Europe with greater frequency and devastation. The plague of Orosius arrived in Rome in 125. Like several predecessors, it had started in Africa, in the wake of a famine caused by locusts. Chroniclers said the infection had killed 800,000 people in Numidia and more than 200,000 elsewhere in North Africa. The full toll in Europe is unknown, but at least 30,000 Roman soldiers died. As terrible as it was, this epidemic, perhaps measles, would seem only a foretaste of the horror when the plague of Antoninus began. Its arrival may have marked the first time a new plague reached Europe from Asia.

Rome's empire now touched India and the steppes leading to Mongolia and China. China's empire had as many people as Rome's, some 50 million, and it was adapting to germs that were still deadly strangers to the West. In the second century, European and Asian pathogens met through three new links: ships, caravans, and the Huns.

Western seamen had found routes across the Indian Ocean to the South China Sea, and they traded directly with the Han court in imperial China. Their ships could carry infected men, insects, rodents, and cargo; pathogens could breed in their water and provisions. There was East–West contact also through caravans along the Silk Road; the Chinese exported raw silk to the Roman province of Syria, where it was rewoven and traded westward. Trade by land and sea thus connected peoples from the Atlantic to the Pacific, and could carry diseases traveling both east and west.

Just as important in spreading disease were the Huns. Late in the first century A.D., these ferocious nomads came riding out of their homeland northwest of China. They migrated through the steppes of Asia and continued westward, stampeding terrified tribes before them. For centuries, waves of displaced people, followed by conquering Huns, would bring wars and epidemics to the Roman world, eroding its borders and thinning its ranks. One of the earliest visitations was the plague of Antoninus, the worst pestilence Rome had ever known. The Romans first met it halfway to China, in Syria.

In 164, Roman troops were sent to Syria to quell a local revolt. They fought there for two years, dying of both battle wounds and some previously unknown contagion. When the legions returned home in 166, they carried the infection with them, and it spread like fire to every corner of the empire. It raged for fourteen years, killing from one-quarter to one-third of Italy's population and, in all of Europe, some 4 to 7 million people. One of its last victims was the philosopher emperor Marcus Aurelius, who succumbed to it in Vienna in 180. The epidemic returned briefly in a milder form nine years later. At its height, it had killed 2,000 people a day in the city of Rome alone.

This epidemic is also called the plague of Galen, after the physician who described it. Centuries later, when Rhazes read Galen's account, he felt sure he recognized a disease he knew well, smallpox. Galen did not mention pockmarked survivors, but otherwise his description matches later ones of smallpox striking virgin populations. The diagnosis also fits our knowledge of the early history of the disease. Although there are some dissenting views, smallpox probably began in India several thousand years ago and was carried by the Huns east to China and west to Europe, where it wrought utter havoc.

As bad as it was, the plague of Antoninus turned out to be neither Rome's last nor its worst. In the year 250 came the epidemic that brought St. Cyprian so many terrified converts. It had arisen in Ethiopia, wrote Cyprian, and reached Carthage in 252. It continued to travel quickly, spread person to person and by victims' clothing. The symptoms included violent vomiting and diarrhea, high fever, skin lesions, sore throat, and sometimes gangrene of the hands and feet. Again the description suggests smallpox or a similar crowd disease mauling a virgin population. People panicked and fled cities for the countryside, but the disease followed them, wiping out refugees and whole villages of farmers and craftsmen. The epidemic recurred cyclically for sixteen years and killed more than half of its victims. At its peak, more than 5,000 people died in Rome each day.

This disease, or something like it, returned on and off in Europe for another three centuries. St. Bede, in England, described its occurrence there in 444; it was so severe that in some places there were hardly enough survivors to bury the dead. In 580 in France, Gregory of Tours described a dreadful epidemic, probably of smallpox. Historians argue about whether some of these epidemics were smallpox, measles, typhus, or even isolated instances of bubonic plague. What is certain is that wherever there were populations dense enough to support crowd diseases, spasms of infection were beating Europe to its knees.

From the second century through the fifth, famine, migration, war, and pestilence followed each other in a dismal round. Soldiers, traders, and wandering tribes kept bringing new diseases to Europe, with staggering death tolls. Land deforested for shipbuilding eroded, farms declined, urban hygiene deteriorated; the social fabric was rent by administrative failure, economic decay, and rebellions. Eventually Europe was a scene of abandoned villages and shrinking cities. Its crafts and industries faded, and literacy dwindled. The Roman world was sliding back into the rural life of an earlier era, its physical and social health broken. The bones of Roman peasant farmers from the late years of the empire show signs of chronic anemia, poor nutrition, and stress fractures—the marks of poverty, overwork, and disease.

It was no consolation that Rome's invaders suffered as well. The Huns drove before them Alans, Vandals, and Goths; they poured into Italy, France, Spain, and North Africa, carrying infections from their own lands and perhaps from Asia. But as they overran Europe's towns and villages, they must have met as many unfamiliar diseases as they carried. In 452, when the Huns reached the gates of Rome, they halted and then fell back. Some people credit the persuasive power of Pope Leo I, others a miraculous intervention of God. One should also credit the epidemic—perhaps smallpox—that raged within the city's walls and among the besieging Huns. In any case, the prize of Rome was becoming more symbolic than material. It would soon be reduced to a ruin in a malarial waste, holding a mere few thousand people.

As Rome staggered through its long sickness toward dissolution, these waves of epidemics attacked the world's other urbanized regions. Their crises, however, did not happen at exactly the same time or in quite the same way. Each region had a distinctive history, environment, and degree of adaptation to crowd diseases.

The Middle East, the world's most ancient center of city life and crowd infections, probably had reached relative disease stability by Roman times, its once lethal diseases having subsided to childhood ills or endemics. Yet the area did have to cope with new contagions from east and west. The situation was probably worse in India and China; like Europe, they were epidemiologically less stable than the Middle East, still adapting to a population explosion and a perturbed environment. India's borders made it vulnerable to invasion from the north, so it often exchanged pathogens with raiders and conquerors. Although the details are sketchy, India, like Rome, probably had its population cut back sharply by pandemics, especially measles and smallpox. China also grew swiftly in population and social organization, only to suffer a sharp die-off. From 200 B.C. to A.D. 200, its population declined severely because of new epidemics. Both China and Japan were ravaged by smallpox for centuries, and some places had mortality rates as high as 50 percent.

It is significant that Europe, India, and China had different disease patterns between north and south. In Europe, the Mediterranean suffered more disease than the north, probably because of its bigger cities, its busier ports, and the climate's ability to support subtropical germs and vectors. In India, the longest-settled region was the Indus valley, in the north, with its dry, warm climate. As a swelling population required more food, settlement expanded into the hot, wet Ganges valley. The south was ridden with malaria, dengue, perhaps cholera, and eventually urban crowd diseases. To this day it remains densely populated, poor, and less healthful than the north.

A similar picture of development and disease patterns appeared in China. Its empire originated in the north, in the Yellow River valley. The Yangtze valley, in the south, was hot and humid, as different from the north as Florida is from Vermont, or Sicily from Denmark. Subduing the south for farming sent millions of peasants into rice paddies, where they lived hard, short lives, many of them afflicted with malaria and schistosomiasis. The environment and disease pool in the south did not stabilize until centuries after they had in the north.

Thus the semitropical south of both India and China developed later than the temperate north, and differently. A similar lag in development occurred in sub-Saharan Africa and much of Southeast Asia. For millennia the people of these areas would suffer ill health and export infections. Temperate urban areas would suffer in their own way; though spared debilitating tropical parasitisms, their dense populations would be linked in catastrophic epidemics. The coup de grâce that set them all back a thousand years was the plague of Justinian.

In the fourth century, the Roman empire had split in half, with capitals in Rome and Constantinople. In the next century, Rome was sacked by Goths and Vandals; the western empire crashed before the century ended. The eastern realm survived; as the middle of the sixth century neared, the eastern emperor, Justinian, was reconquering western territories from German tribes. By 542, he had taken back much of North Africa, Sicily, and part of Spain, and was poised to go after more. Then came the plague.

While the word "plague" is used for any severe epidemic, what hit Constantinople was not *a* plague, it was *the* plague, bubonic, which in a later outbreak would be called the Black Death. This may not have been bubonic plague's first appearance; some historians think that it attacked Philistines at Ashdod, and that the plague of Antoninus was bubonic, as was a terrible epidemic of 200 B.C. in Egypt and Libya. But only the pandemic of Justinian's reign is indisputably bubonic plague, and it caused one of the worst die-offs in human history.

The great chronicler of this disaster was the Byzantine historian Procopius, who described "a pestilence by which the whole human race came near to being annihilated." The outbreak, he said, started in lower Egypt in 540. It traveled down the Nile to the port of Pelusium in 542 and spread by ship to Constantinople. Procopius' account of its ravages is hair-raising.

The sickness began with fever. On the first or second day, buboes, or swollen lymph glands, appeared in the armpits, groin, and neck. Fever raged, agonizing buboes swelled, and the germ assaulted the nervous system, causing lethargy or hallucinations. Half or more of those infected died by the fifth day. When cold weather arrived, the disease changed to its pneumonic form, which was spread by coughing and destroyed the lungs. People died vomiting blood and choking on it. Pneumonic plague was—and without antibiotics still is—fatal to 95 percent of its victims.

Panic, disorder, and murder reigned in the streets of Constantinople. There were too many corpses to bury. The roofs were removed from the city's fortified towers, and bodies were stacked in them like cordwood. Soon the towers filled, and the stench became unbearable. People kept dying, up to 10,000 each day, and there was no place to put the corpses. Rafts were loaded with the dead, rowed out to sea, and set adrift. When this bout of plague ended, 40 percent of the city's people had died.

The plague spread rapidly to coastal cities all over the Mediterranean and traveled more slowly inland. For six straight years it devastated Italy, Spain, France, the Rhine valley, Britain, and Denmark. Justinian's attempt to restore the western empire collapsed. Europe's agriculture ebbed, and trade almost halted. Many cities became, like Rome, scourged remnants, with the dead heaped in the streets and the living huddled in churches praying for deliverance.

The plague returned frequently until about 590; it recurred less often, in small, localized outbreaks, for another 150 years. It lasted two centuries in all. Sometimes it was compounded by smallpox, typhoid, and other diseases. When the pandemic ended, Europe's population had been halved. The Mediterranean, more thickly populated than northern Europe, and exposed by trade to more diseases, probably suffered worse. For centuries afterward, it lagged behind the north in population growth and in economic and cultural renewal.

This plague pandemic weakened the Romano-British people and opened the way for Saxon invasion. The Balkans and Near East suffered as badly as Europe. Justinian's realm shrank and was succeeded by the Greek-speaking Byzantine empire. The Middle East and North Africa were ravaged; some historians think Islam marched so swiftly through many lands because the plague had battered them physically, psychologically, and culturally. Plague traveled on to Persia, Arabia, India, and Southeast Asia. It struck China in 610 and recurred for 200 years, doing its worst damage in the south and in coastal provinces. No one knows how many people died; census figures became unreliable or nonexistent as China was convulsed by pestilence and unrest. Perhaps partly because of pandemics, many Asians turned to Buddhism, as many Europeans turned to Christianity during the plague's chaos.

One asks, as with all new diseases, Where did it come from? The conflagration was sparked by ecological changes involving humans and several other species, in a series of unlikely encounters. To see how the plague bacillus traveled from free-living rodents to humans in cities, you must follow the flea in which it travels.

Bubonic plague is caused by the rod-shaped bacterium *Yersinia pestis,* named for Alexander Yersin, the eccentric Swiss scientist who discovered it. *Y. pestis* is a longtime companion of burrowing rodents such as field mice, ground squirrels, and marmots. It is transmitted by the rodents' fleas, which drink it in the host's blood. *Y. pestis* multiplies in the flea's gut until it forms a mass that causes blockage and, when the flea feeds, regurgitation of germs. Flea bites carry the germ from rodent to rodent. This ancient rodent-flea-bacillus connection usually causes only limited, periodic outbreaks of disease.

In its familiar hosts, bubonic plague flares up in cycles of two to twenty years; the reason may be that changes in weather and food supply allow a population increase, and crowding somehow induces flare-ups. The disease does seem to act as a brake on overpopulation, although it is not invariably fatal to its natural hosts. But when *Y. pestis,* having smoldered among burrowing rodents, reaches unfamiliar species, it is one of nature's deadliest bacteria. A rodent unaccustomed to *Y. pestis* can come into contact with one that is, and pick up its germ-laden fleas. If, for instance, a rat catches plague from the flea of a ground squirrel, the rat's fleas then acquire *Y. pestis* from their host. When the rat dies of plague, its fleas leave the body to seek a new home. Thus a deadly epidemic is set off among rats. If humans are nearby, the fleas bite them, and catastrophe may ensue. Clearly, that is the end result of an unlikely series of chance encounters. The risk of epidemic human plague arose only when farms and cities began attracting large numbers of scavengers, especially the black rat.

Knowing all this, we can reconstruct the steps leading to the plague of Justinian. The disease started millions of years ago in rodents in the Himalayan foothills, between India and China. It spread to wild burrowing rodents in China, the Middle East, and eastern and northern Africa. It still had little importance to humans; they met it only sporadically, when a hunter killed an infected animal. Some 2,000 years ago, an unknown environmental change, perhaps weather fluctuations, raised rodents' food supplies and boosted their population. Predictably, *Y. pestis* infection broke out. This had happened countless times before, but now human settlements had sprung up, and they attracted the Indian black rat. This hardy, adaptable creature is a weed species, one that thrives in disrupted environments. It flourished in people's barns, alleys, and ships.

Black rats were intermediaries between infected wild rodents and humans. They picked up *Y. pestis* from the wild rodents, and they infected their own fleas. The most common rat flea, *Xenopsylla cheopis,* is as hardy and opportunistic as its host; it always deserts a sinking rat. As rats died of plague, the infected fleas found their way to humans. Soon people began to die as quickly and miserably as rats.

At first, people could not transmit plague, but soon two forces amplified the disease, one natural and one manmade—cold weather and ships. Cold weather shifted the attack of *Y. pestis* from lymph nodes to the lungs; the pneumonic form of plague could spread directly from person to person, by coughing. And ships carried plague-infected rats far and wide. The black rat, a nimble climber, could scoot up mooring ropes; as a result, it was carried from India to the eastern Mediterranean and eastern Africa. From Egypt, the rat and plague went by ship to Constantinople and then to the ports of Europe. Plague also traveled eastward from India, to southern China and Japan.

Scholars debate some points in this scenario. Some say bubonic plague originated in Africa or in both Africa and Asia; others say it arose in Asia, and reached Egypt primarily by land, over the Silk Road. Nevertheless, there is basic agreement on the confluence of circumstances: environmental fluctuations, an excess of plague-infected wild rodents, their contact with rats scavenging human homes and ships, the ability of some fleas to feed on several hosts, the ability of the disease to shift to human transmission. Thomas Hardy never invented such a web of coincidence. But given the complexity of ecological interplay, and the degree to which people had changed their surroundings, there were bound to be some unlikely, dramatic consequences.

Near the end of the pandemic, plague attacked less often and less violently. The final episodes were isolated outbreaks in Italian and Spanish port cities, perhaps reintroductions of *Y. pestis* from ancient endemic centers. Then the germ retreated to its usual quiet home in wild rodents, though now the dormant infection was spread over a larger area than before.

To this day, no one knows why the plague of Justinian faded. Some people think that the bacillus mutated to a more benign form, but to others this seems far-fetched. There have been three known plague pandemics, and after each one, the microbe subsided to its old animal reservoir, far from humans. But after each attack, *Y. pestis* dwelled in more of the world than before. Plague is still a disaster awaiting an opportunity, as the 1994 outbreak in India grimly proved.

When the first pandemic receded, Europe felt it had been reprieved from death. So did much of the rest of the world, from North Africa to Japan. Europe's population had been cut in half, and city life virtually ended. Cured, for the moment, of diseases bred by urban growth, Europe could recuperate. That required a centuries-long rest which would be called the Dark Ages.

Six

the flying corpses
of kaffa

*A rural convalescence. Sweats and
lepers vanish; bacteria compete.
Disasters of the Little Ice Age.
Horrid acts at Kaffa? Mongols, a new
breed of rat, and the Great Dying.*

Many people think of the Middle Ages as a dark pit of squalor and
sickness. Actually, it was a more healthful, less violent era than those
before and after. Bubonic plague, having halved the population of some
parts of the world, ebbed. In the eighth century, it disappeared entirely;
the chain of infection had broken in both humans and rats. Events
conspired to create a time of healing. The Dark Ages, from the fall of
Rome to about the year 1000, gave both humanity and the land a rest
they desperately needed.

Fossil pollen, tree growth rings, and other indicators of past climates show that around the year 800, Europe entered more than four centuries of warm weather. Mild winters and dry summers allowed farming to flourish, the food supply to rise. Because of the smaller population and its reduced activity, forests were no longer being felled for shipbuilding and new farmland, and exhausted fields went fallow. In much of the Old World, ecological climax states returned. Meanwhile, the tribes that had carried war and pestilence across Eurasia were settling down. Industry, trade, and travel nearly vanished; a shrunken population lay dispersed in farms and villages. For many centuries, the disease pool remained stable. The consequences of the great die-off offered a reprieve from contagion.

Since the dawn of the Neolithic, 8,000 years earlier, virulent new diseases had arrived with increasing frequency. Now the need to adapt to new infections almost vanished. The killer epidemic diseases of the Roman era began subsiding to routine childhood illnesses. The bones of Saxons, Franks, and other early medieval peoples show arthritis and stress fractures, the stigmata of farming's repetitive drudgery, but little malnutrition or infection. Even the toll of violence dwindled; medieval skeletons, except for those of Vikings and some scattered tribes, show relatively few combat traumas. Life was hard in the Dark Ages, and shorter than it had been for Roman city dwellers, but it was lived without recurrent wars, famines, and plagues.

The results of this recovery were a rising population and growing prosperity. By the end of the first millennium A.D., Europe began to stir from its rural convalescence. Thanks to improved farming methods, good weather, and above all lack of epidemics, Europe's population more than doubled between 1000 and 1300, reaching its highest point in a thousand years. England became an exporter of wheat and rebuilt its roads, which had fallen into disrepair since Roman times. A similar recovery occurred on much of the Continent. Walled cities reappeared, some with as many as 100,000 inhabitants. There was a revival of crafts, trade, and travel. The Crusades sent crowds from all over Europe to the Near East and back; although they caught local diseases in the Near East, they ignited no major epidemics when they returned. This was one more sign that the major pools of infection in Europe, North Africa, and the Near East had more or less merged and now held few surprises for each other.

Europe had its familiar old infections, and the intestinal ones that nag all small communities with poor hygiene. Gonorrhea, which probably existed in cities of the ancient Middle East, also must have been familiar in the medieval cities of Europe and North Africa. Malaria extended from the Mediterranean as far north as Scandinavia, its incidence and severity varying with such local factors as water salinity and the number of cattle to serve as alternative hosts for mosquitoes. But to our knowledge, only three serious diseases appeared or changed in medieval Europe. One was a mysterious malady known as the sweats. It is interesting today chiefly because it vanished.

The sweating sickness was aptly named. It caused fever, flushing, headache, thirst, and cascades of stinking perspiration. Some victims recovered, but many died within a day or two. The sweats first emerged in the early Middle Ages, primarily in Britain, attacking the upper classes more than the poor. Prayers that it might do otherwise backfired. Tradition has it that in 657, two Irish kings, Dermot and Blaithmac, asked an assembly of nobles and clerics to pray for a pestilence to reduce the number of poor people, that the rich might live more richly. Apparently heaven was listening. An epidemic of the sweats promptly killed both kings and many of the priests and nobles.

The sweats either faded or went unrecorded for eight centuries. Severe epidemics occurred in Britain in 1485, 1509, 1517, 1528 (when Germany was afflicted as well), and 1551. After that, the disease disappeared forever. There were, however, outbreaks of a similar, usually milder infection, the Picardy sweats, in eighteenth- and nineteenth-century France. This suggests that the sweating sickness was a familiar endemic in France that took a higher toll when it was sporadically introduced in Britain and other nearby lands. The first epidemic in Tudor England, in 1485, closely followed the arrival of French mercenaries, whom Henry VII had imported to fight against Richard III at Bosworth Field.

Today there is no disease resembling the sweats. It sounds like a viral zoonosis from domestic or wild animals that vanished because the germ or its carrier died out. Few historians give the sweats more than a footnote, but it reminds us that just as new diseases can appear with startling, mysterious abruptness, they can just as quickly and mysteriously vanish. Germs and their carriers, like the victims of infection, are vulnerable to change, death, and even extinction. We will never know how many diseases have failed to sustain themselves and been lost to history. They may have been as numerous as those that survived.

A far more important new disease in medieval Europe was leprosy. Besides having a horrid fascination of its own, leprosy provoked the first recorded debate on whether there are truly new diseases, and if so, where they come from. Furthermore, the history of leprosy is inseparable from that of an even more important disease, tuberculosis; their entwined stories throw light on why diseases wax, wane, and reemerge. The story echoes into the 1990s, when we are seeing a frightening resurgence of tuberculosis.

Leprosy and tuberculosis may seem very different, but the germs causing them are very close cousins. They belong to a group of slender, rod-shaped microbes called mycobacteria, which were free-living in the seas hundreds of millions of years ago. Various types adapted to living in fish, reptiles, birds, pigs, rats, and cattle. Some are harmless to their hosts; others can cause mild or serious illness.

The leprosy and tuberculosis bacilli probably first attacked people less than 10,000 years ago; their original homes were domestic animals or scavengers. The source of leprosy is still in doubt; some scientists think that *Mycobacterium leprae* mutated from an ancestor in mice. British scientist Richard Fiennes believes humans acquired leprosy from the Asian water buffalo, by using its hide. Whatever its origin, leprosy probably arose somewhere in the area from India to East Africa; it was already rife there 4,000 years ago. A thousand years later, it was common in China, and a millennium after that in Alexandria, in Egypt. Alexandria was a metropolis whose palaces cast shadows on hovels crowded with the poor and ill fed. Everywhere, leprosy and tuberculosis thrive on social and physical misery.

Leprosy behaves like no other infection. It has adapted uniquely to humans and now occurs naturally in us alone. To study it, researchers must artificially infect armadillos or the footpads of mice. Although leprosy inspires terror of contagion, it is perhaps the most difficult disease to transmit. One can share home and bed with a leper for decades and not catch it. The disease can take years or decades to develop, and as long again to do its worst damage. Leprosy does eventually disfigure and injure the body; the classic symptoms are leonine swelling of the face, skin lesions, weakness, paralysis, and mutilated fingers and toes.

Some diseases are more repulsive and far deadlier, but none has inspired the particular horror of leprosy. Objects of fear and loathing, lepers have been killed, driven from society, or forced to wear hoods, badges, and bells to warn off the rest of humanity. Attempting to erase the stigma, scientists renamed leprosy Hansen's disease, for the Norwegian doctor who discovered its cause. Although cure is now possible, science has expunged neither the old name nor its evil magic.

Unfortunately, that name was long used for many people who did not have leprosy at all. Leprosy probably existed in the ancient Near East, but the use of the word in English versions of the Old Testament is a mistranslation. The original Hebrew word, *tsara'at,* suggested ritual uncleanness more than any one medical condition. It seems to have signified any unsightly skin infection, such as ringworm, eczema, leprosy, or syphilis. When the Old Testament was translated into Greek in the third century B.C., *tsara'at* was rendered as *lepra,* which was used similarly in Greece for disfiguring skin diseases. At that time, *M. leprae* probably had not yet reached Europe.

Five hundred years later, leprosy was appearing for the first time in Greece and Rome; Roman legions may have spread it throughout the empire. Medical writers wondered how Hippocrates had missed a disease so unmistakable, so serious, and now rather common. Some thought it was a recent arrival from Egypt, while others said it was a new disease. The idea of a new disease set doctors and philosophers into fits of dispute.

Plutarch's *Quaestiones conviviales* contains his discussion with friends about whether new diseases could come into existence. Some said that changes did happen in nature, so new diseases might arise. Others argued that every disease had always existed; many ills had merely gone unnoticed by doctors. The latter view made more sense in a time when diseases were thought to result from an imbalance of humors and the effects of diet and climate. By the end of the debate, most of Plutarch's friends agreed that nature created no novelties; changes in diet and in bathing habits could account for all apparently new illnesses. Plutarch mentioned but rejected Democritus' theory that extraterrestrial atoms might cause unusual epidemics, an idea that would be revived from time to time until today, as the panspermia theory.

The oldest physical proof of leprosy's existence in the West is in Egyptian skeletons with telltale pathology, dated to around the year 500 A.D. By then, the disease may have already spread to much of southern Europe and even beyond. As the Dark Ages passed, leprosy became more common, traveling as far north as Britain and Sweden. Special hospitals called lazar houses (for St. Lazarus, the patron saint of lepers) were built to keep lepers away from society. By the thirteenth century, there were thousands of lazar houses in Europe, and some 200 in England alone. More than ever, leprosy was seen as divine punishment, a curse marking the moral outcast.

The epidemic peaked in Europe around 1300. A century later, leprosy had almost vanished there except in Scandinavia and a few other scattered areas. Historians have puzzled over its disappearance in Europe, not least because it continued to flourish elsewhere; it still afflicts 15 million people around the world, despite the availability of curative drugs. Various theories are that lepers, weakened by their illness, were wiped out by the Black Death; that improved nutrition defeated leprosy; that the germ mutated and became less infective. None of these ideas explains why the disease faded in most parts of Europe but persisted in others. A better answer came with understanding that although the symptoms of leprosy and tuberculosis are different, the germs that cause them are similar.

There are two main types of tuberculosis bacilli, *Mycobacterium bovis* and *Mycobacterium tuberculosis*. The former has long lived in wild cattle, causing few symptoms or none. When people domesticated cattle, lived with them, and consumed dairy products, they acquired *M. bovis*. The germ might make its home in the human bowel, the lymph glands of the neck (appearing as scrofula), or the spine (as Pott's disease). The angular hump peculiar to Pott's disease has been found in skeletons from Europe's Neolithic and Bronze ages, up to 7,000 years ago, and in Egyptian mummies.

M. bovis infection was nasty enough, but a worse type of tuberculosis appeared some 4,000 years ago. *M. bovis* gave rise—by quick mutation, slow adaptation, or both—to *M. tuberculosis*. This bacillus thrives in the oxygen-rich tissue of human lungs; it can be transmitted from person to person by coughing or close contact, especially in crowded environments. Humans also gave it to their herds, which reinfected them in turn. *M. tuberculosis* did its worst damage among the crowded urban poor, preying especially on the young, ill fed, and overworked who lived without sunlight and fresh air.

Consumption, or pulmonary tuberculosis, soon became widespread in the cities of Asia and the Near East. It spread to Greece, where

Hippocrates described it in detail. It was so common in late Roman times that it survived for a while even in the sunlight and fresh air of early medieval village life. As the more healthful rural conditions of the Dark Ages continued, consumption waned.

Only recently was it noticed that leprosy in Europe increased as tuberculosis faded, and that a thousand years later, as leprosy faded, tuberculosis returned. This has special significance in light of the discovery that *M. leprae* and *M. tuberculosis* are so closely related that in laboratory tests, each causes an immune reaction in the presence of the other. The logical conclusion is that within the human body, the germs cross-immunize; each confers resistance to the other.

Cross-immunity between leprosy and tuberculosis is not automatic and absolute, but the presence of either disease does greatly lower the odds of catching the other. Of the two germs, *M. tuberculosis* is a more efficient preventive than *M. leprae*, because it is more aggressive; it is more contagious and strikes earlier in life. As Europe's urban population grew in the late Middle Ages, many young children were exposed to *M. tuberculosis,* had mild symptoms or none, and afterward were resistant to *M. leprae.* Thus tuberculosis won an edge over leprosy in the struggle to find a niche in the human body. It is a case of ecological competition, with *M. tuberculosis* winning.

Just as the ruralizing Dark Ages gave an edge to *M. leprae,* the urbanizing late Middle Ages offered an advantage to *M. tuberculosis.* The changing balance shows that even small differences in a germ, in a host, or in a natural or manmade environment can alter the pattern of human disease. This ecological view of competing pathogens will later help explain the waning and resurgence of other infections in the age of AIDS.

For a while, the ascendance of tuberculosis over leprosy was the only major change in Europe's disease pool. The population, and its concentration in cities, continued to grow. Revived sea and land routes connected Sicily and Sweden, the Baltic and the Balkans. The Crusades put Europe in contact with the Middle East, which had resumed trade with eastern Asia; Europe soon found itself in direct contact with China again. Through Arab middlemen, it traded with the interior of Africa for gold. By 1300, an elaborate network crisscrossed Eurasia, as it had in Roman times, allowing trade in grain, metal, furs, honey, silk, and spices. New cultural contacts and increasing literacy, especially in southern Europe's castles and cities, promoted what would be called the Renaissance.

Then, in the middle of the fourteenth century, came the worst disaster in human history, the second bubonic plague pandemic, the Black Death. It had the usual precursor, a Malthusian crisis of rising population, strained resources, and environmental change. Part of this prelude was Europe's rising reliance on single-crop wheat farming. This exhausted the soil and created a threat of famine when wheat harvests failed. From about 1250 on, cycles of famine did occur as the climate changed. By 1300, what historians call Europe's Little Ice Age was under way. Its cold, wet weather would last 500 years.

The 1290s brought especially heavy rains, crop failures, and famines. Another cold, wet period began in 1309, creating the worst decade of famine in Europe's history; crops rotted in the fields, livestock epidemics depleted herds, and people were reduced to eating cats and dogs. In many cities, 10 to 15 percent of the population starved to death. Cold, wet summers, wheat failures, and famine continued until 1325, and recurred several times for another two decades. A hungry people are a sick people; intestinal diseases were rife. By the 1340s, Europe's population was declining, and the earth itself seemed to rumble with anger. There was a series of natural disasters, from plagues of locusts and mice to floods, earthquakes, and a major eruption of Mount Etna. To make things even worse, the bloody Hundred Years War was being fought between England and France.

Climatic change and natural disasters were not limited to Europe. In 1333, China was hit by drought and famine; then came storms, earthquakes, floods, locusts, and epidemics. Revolt broke out against the Mongols, who had invaded and occupied China; the uprising was complicated by civil war. The rest of Asia did not escape upheaval. In the 1330s, unusually dry, windy weather sent Mongol and Turkic nomads migrating into new territories, in search of food and water. The weather roused rodents bearing *Y. pestis* from their burrows in central Asia. As they searched for better homes, they infected more susceptible rodents. Soon those rodents and Mongol horsemen carried *Y. pestis* across Asia. The disease now extended from Manchuria to the Ukraine.

Bubonic plague, we saw, is not normally a disease of humans. Each pandemic has resulted from a biological accident in which *Y. pestis* escaped its old relationship with wild burrowing rodents and their fleas, to attack other rodents and, through them, humans. In the fourteenth century, as in Justinian's day, the chief intermediary between wild rodent and man was the black rat. This native of India had come to depend on the food and garbage of human settlements. Originally a tree dweller unaccustomed to *Y. pestis*, it was adept at climbing mooring ropes and scrambling to the upper levels of barns and homes. It was brought to Europe by the ships of returning Crusaders, and by the fourteenth century had replaced indigenous rats there. As a disease carrier, the thriving black rat was like kindling awaiting a spark. The conflagration was started by Genoese traders liberated from a Tatar siege in 1346.

In the 1330s, plague had begun to spread from central Asia to China, India, and the Middle East. By 1340, Arab chroniclers reported that people were dying in droves from Tatary to Armenia. In 1346, plague had followed caravan routes west to the Crimean port of Kaffa (now Feodosiya), on the Black Sea. The city had been under siege for three years by Janibeg, khan of the Kipchak Tatars. Genoese traders who had taken refuge within Kaffa's walls remained trapped there. When plague started killing the Tatars, Janibeg had to withdraw with his surviving troops and leave behind heaps of unburied corpses. According to what may be either history or legend, he took a deadly parting shot: he used his catapults to propel plague-infected corpses over Kaffa's walls. The besieged people cast the bodies back over the walls, into the sea, but the seeds of plague had entered Kaffa.

Perhaps flying corpses really did bring plague to Kaffa. Perhaps scavenging rodents were responsible; scholars who mistrust the dramatic on principle prefer this explanation. Whatever the case, the retreat of plague-ridden Tatars left the Genoese traders free to head for home. In the summer of 1347, they hurried to their ships and sailed for Italy. They must have carried infected rats on board, for plague soon broke out in the Mediterranean ports they touched. This was not the only introduction of *Y. pestis* into Europe, but it may have been the first of that pandemic.

Merchants from Genoa and Venice carried plague to other ports of Italy. From these it went as fast as ships could travel to other Mediterranean ports, then on to towns and cities on the Atlantic coast and along Europe's main rivers. People fled port cities, but many succeeded only in dying of plague in roadside ditches. By late 1347, when the weather turned cold, bubonic plague had blanketed most of southern Europe. It took its deadlier, pneumonic form and traveled even faster. It reached England in 1348. At the same time, it was racing eastward from Asia toward Moscow by land, and by ship to the Persian Gulf, the Arabian Peninsula, and the Nile delta.

The symptoms Procopius had described eight centuries earlier were repeated in Boccaccio's *Decameron,* a set of tales told by Florentines secluded in the countryside to escape the plague. Buboes appeared in the groin and armpits, swelled to the size of small apples, and spread all over the body; when black and purple spots appeared on the skin, death was certain. In the pneumonic and septicemic (blood-poisoning) forms of plague, people might get up healthy in the morning and die before night, raving and vomiting blood. Half or more of the bubonic victims died, as did almost all of the pneumonic and septicemic ones.

Not knowing the cause of plague, people could only guess how to cure it. They tried purging, bleeding, fumigating, cauterizing buboes, and bathing in their urine. The Venetians invented the quarantine, isolating arriving ships for forty days; since that confined sailors but not ship rats, it failed to contain plague. Some people did notice that when plague arrived, dying rats staggered from their hiding places, but no one made the connection to human contagion.

The idea of contagion, in fact, did not yet exist. People blamed the plague on miasmas, earthquakes, comets, cats, dogs, lepers, Gypsies, and especially Jews. Pogroms began all over Europe; in Mainz alone, 12,000 Jews were burned alive. A more benign theory came from the great surgeon Guy de Chauliac, who thought one could catch plague merely by looking at someone who had it. The medical faculty of the University of Paris told Pope Clement VI that the reason for the plague was a conjunction of Saturn, Mars, and Jupiter in the house of Aquarius, which had occurred on March 20, 1345.

As always, pestilence was seen as punishment. Only the most dreadful individual and collective sins could account for so much dying. Flagellants by the tens of thousands whipped themselves through the streets of Europe, trying to appease divine wrath with their repentance. At first the pope blessed their processions, floggings, and cries of *"Mea maxima culpa."* When they became too numerous and rebellious, virtually a church unto themselves, he suppressed them with sword and fire. The plague, of course, continued.

Petrarch, who lost his beloved Laura to plague, thought posterity would think the testimony of his age mere fable. Later generations would not be able to imagine the empty houses, the abandoned towns, the suffering, madness, and death. If people who lived it could hardly believe it, why should anyone who had not? Chronicles and diaries of the plague years still challenge the imagination. There are stories of infants sucking at the breasts of their dead mothers, lone survivors of entire towns walking through empty manors in robes and jewels, naked orgies in the streets, ghost ships manned only by corpses adrift at sea. In Siena, one survivor wrote:

> Father abandoned child, wife husband, one brother another . . .
> none could be found to bury the dead for money or friendship . . .
> they died by the hundreds, both day and night, and all were
> thrown in ditches and covered with earth. And as soon as those
> ditches were filled, more were dug. And I, Agnolo di Tura . . .
> buried my five children with my own hands.

When Clement VI asked to know the toll, he was told 42 million dead, 25 million of them in Europe. It is impossible to know how much to credit medieval mortality figures. Some historical revisionists of recent decades have minimized the plague's impact and lowered estimates of deaths. Surely some of the numbers were exaggerated, but recent studies of city records from all over Europe show that the dying was as ghastly as many chroniclers claimed. A few cities, such as Milan, went almost untouched, but even by conservative estimate, at least one-third, and perhaps as many as one-half, of the people of Italy died, and probably the same proportions in France and England. Russia, Poland, and the Balkans suffered just as badly. At Avignon, when no burial ground remained unused, the pope blessed the Rhone so that the corpses dumped into it would have a consecrated home. Vast numbers of people died in London and Rome, and all but five individuals in Smolensk. Europe lost between one-quarter and one-half of its population.

The toll was at least as dreadful in the rest of the Old World. The Byzantine empire, staggered by plague, collapsed. North Africa was ravaged, and the Islamic world lost one-third to one-half of its people.

India suffered in similar measure. In China, between 1200 (before Mongols and plague arrived) and 1400 (when the Mongols were gone and plague was ebbing), the population dropped by half, to 65 million. The Mongols' empire dissolved; their centuries of raiding and conquest, from China to Europe, had ended, and they drifted into the margins of history.

Another half-dozen major waves of plague struck Europe before the fourteenth century ended. It returned in many places for 200 years or more, sometimes compounded by smallpox, typhus, malaria, and dysentery. Until the seventeenth century, it remained common for plague to wipe out as many as half of a city's people. The best-known recurrence was England's last, the Great Plague of 1665, which Daniel Defoe described in his *Journal of the Plague Year,* and Samuel Pepys chronicled in his diary. The final outbreak in France came in 1720–1721, in Marseilles. The disease returned occasionally outside Europe, in such places as Egypt and Southeast Asia, into the twentieth century. The third pandemic, at the turn of this century, made plague endemic in the wild rodents of the western United States.

The phrase "Black Death" was first used two centuries after the second plague pandemic began; until then, Europeans called it the Great Dying. It was a disaster without equal in human history. Somewhere between one-quarter and one-half of the people in Europe, North Africa, and parts of Asia perished, and populations continued to decline into the second half of the fifteenth century. It is little wonder that no aspect of European life, from its art to its commerce, was untouched.

The West's artistic imagination would be haunted for centuries by images of the Dance of Death and of St. Sebastian, a patron saint of plague victims. The post-plague labor shortage eroded the landed nobility's wealth and power, and turned serfs into tenants, landowners, or independent artisans. Eventually it stimulated European involvement in the slave trade. Social dislocations spurred uprisings of peasants and workers. The Church's inadequate response to bubonic plague weakened its hold on people's trust and faith; this helped secular leaders expand their authority. The Great Dying did not end feudalism or, by itself, bring about the Renaissance, the Reformation, or the rise of the secular state, but it did hasten their arrival, and many other changes. Europe's center of trade and prosperity shifted from the more severely ravaged Mediterranean to the resource-rich northeast. The paranoid demonizing of Jews as poisoners and plague carriers started an exodus that moved most of European Jewry from the west of the continent to the east.

Plague also probably retarded Europe's settlement of the New World. Before the pandemic, Norse colonies in Greenland had made sporadic contact with Vinland, as Norsemen called the North American coast, but the Greenlanders succumbed to plague, other epidemics, and the weather of the Little Ice Age. Greenland ended up empty and forgotten; it had to be rediscovered by an Englishman in 1585. Thus the Black Death broke Europe's fragile link with North America. Had it remained intact, the New World might have been explored and settled a century earlier.

This second pandemic did even more damage in Europe than the first, yet the results there were less severe. The West did not retreat into a dark age of convalescence, as it had after the plague of Justinian. Instead, the years 1350–1500 were a period of complex transition, in which Europe gathered itself for a huge leap forward. There followed an age of global exploration, and a change in world disease patterns that dwarfed any in the past.

the deadliest
weapon

*An awful restlessness. From the Canaries
to Cuzco. The white man's grave,
the black man's value: disease breeds
slavery. Poxed at Oxford. Siberians,
Maoris, and a global germ pool.*

In 1992, much of Europe and the Americans celebrated the quincenten-
ary of Columbus' arrival in the New World. There were scattered
demurrals and protests. Some people asked, A world new to whom?
Certainly not to Native Americans, some of whom boycotted or disrupted festivities. Other people pointed to Viking landings in North
America centuries before Columbus. Zealous multiculturalists, hostile
to anyone European, white, male, dead, and possibly even alive, condemned Columbus as a fanatic, ignoramus, racist, and exploiter: he and
his like should be scorned, not celebrated.

Some of the arguments have merit, but Genoa, New York, and Buenos Aires were not festooned with black. No matter how one judges Columbus or the fruit of his labor, his voyage demands commemoration. In fact, the better one grasps its consequences, the more awesome and compelling they seem. His voyage and Magellan's circling of the globe were lightning strokes in an unprecedented storm of exploration. They forever changed human health and history, and they altered the world's ecosystem, from flora and fauna to microbes. In one century, Europe launched social and biological revolutions that previously would have taken thousands or millions of years.

That the blaze of exploration should happen was unlikely, and no list of reasons fully explains it. In fact, earlier history would suggest a quite different course. From 1000 to 1300, population and urbanization grew in western Europe. As is typical of swelling populations, Europeans engaged in more trade, territorial expansion, and war—the Crusades, campaigns against Moors and Turks, and the Hundred Years War. Then the Little Ice Age, famine, and bubonic plague brought a die-off that lasted half a century.

By the fifteenth century, Africa and Asia also were reeling from plague and natural disasters. They reacted as one would expect; after sinking sharply in population and prosperity, they slowly began to recover. Their history in the late Middle Ages reflects a struggle for survival and stabilization, not growth and change. The Chinese turned inward, trying to ignore the world beyond their borders. The high Arab culture of pre-plague times had begun to decline, and Muslim armies no longer drove north and west. Soon the Arabs would be driven from Spain, and the Turks checked at Europe's Balkan threshold.

Europe, too, was traumatized, depopulated, and wracked by epidemics, yet it took a different course. It seemed seized by a great restlessness. The fifteenth and sixteenth centuries saw social and religious ferment, economic growth, and dramatic advances in mining, metallurgy, printing, timekeeping, shipbuilding, navigation, and weaponry. Crafts, commerce, and urban life were all revitalized. A search started for new trade routes and for permanent colonies beyond the seas.

Europe's aggressive drive and huge success in this era have no clear explanation, only analogies. They evoke the equally inexplicable surge of ambition that made an Italian tribe called Latins the masters of an empire on three continents. But Europe's post-plague expansion drew on technology the Romans lacked, and since the world's other powers offered little competition, it went unhindered. Europe exploited its advantage with imagination, daring, and ruthlessness. No one dreamed that its greatest impact would be a new world ecology—a stage for the increased emergence and transmission of epidemic diseases.

Early in the fifteenth century, in an overture to the new era, Iberians explored and exploited three chains of islands off Africa's western bulge, the Azores, Madeiras, and Canaries. The Azores and Madeiras were uninhabited, so it was easy for the Portuguese to burn the native forests and introduce wheat, grapes, goats, rabbits, camels, honeybees, rats, weeds, and a variety of insects and microbes. They were so successful in creating a biological neo-Europe, complete with many of its scavengers, pests, and parasites, that today we have no idea what these islands originally looked like.

The most important Portuguese transplant in the Madeiras was an Asian import, sugar cane. Since most of Europe depended on honey as a sweetener, this new crop offered a commercial jackpot. But sugar plantations demand intensive labor; that called for the resource post-plague Europe most lacked. The solution was slavery, still an accepted practice in much of the world. Even before the Portuguese turned to the slave trade that already flourished on Africa's west coast, they probably raided the Canary Islands to abduct natives for manpower. These were the Guanches, a people whose fate would be repeated by natives in European colonies the world over.

The Guanches were olive-skinned Caucasians, probably related to the Berbers of North Africa. A hardy Neolithic people, they had been isolated for hundreds or thousands of years from the Mediterranean's crowd diseases. In the fourteenth century, the Portuguese and French began sporadically to raid and trade in the Canaries. In the fifteenth century, Spaniards undertook to subdue the Guanches and turn the islands into vast sugar plantations. The Guanches fought ferociously, but their Stone Age weapons were no match for Spanish guns and horses, and their naive immune systems were helpless before European infections. By the end of the century they were beaten, as much by germs as by firearms.

The Guanches not killed in battle or enslaved on plantations died in epidemics, especially two which the Spanish called *peste* and *modorra*. These may have been any of the common European crowd infections. As usual when virgin hosts meet new microbes, fatalities were enormous. There were originally some 100,000 Guanches. By 1530, they were reduced to a diseased, apathetic handful; by 1600, there remained only some mixed-blood Hispano-Guanches, and soon even this remnant vanished. Today all that survives of them are some mummified nobles hidden in caves and a half-dozen sentences written in their otherwise lost language.

The Guanches were the first victims of what historian Alfred Crosby calls ecological imperialism. He describes the Iberian conquest of the Azores, Madeiras, and Canaries as a pilot program for the reshaping of European colonies in the Americas, Africa, Australia, and Oceania. In all these places, the newcomers would conquer the human populations and Europeanize entire ecosystems. They dared this because they had seen from the Iberian experience in the Canaries that European crops and herds would thrive in all but the most hostile, unfamiliar environments, and that the fiercest indigenes could be beaten despite their superior numbers and home-ground advantage. As a result, tens of millions of natives around the world would die. Conquistadors, and historians who followed their accounts, tended to credit the Europeans' god, guns, and horses, but their deadliest weapon was their germs.

From the Canary Islands, Iberians pushed south along the western coast of Africa. In 1497, Vasco da Gama rounded the Cape of Good Hope, to sail as far as India. In 1519, Ferdinand Magellan's ships set out for South America; they rounded Cape Horn, crossed the Pacific, and returned to Spain in 1522. It had been just thirty years since Columbus' first voyage to the New World. Italian, Dutch, French, and British explorers were extending the Iberian venture. These blue-water mariners, having mastered ocean winds and currents, were free from hugging coasts and island-hopping. They could cross the Atlantic in one trip or traverse the Indian Ocean in weeks instead of months. And wherever they went, even in small numbers, their firearms and infections made them conquerors.

Nowhere were Europe's biological predations worse than in the Americas. The natives had been there far longer than the Guanches had been in the Canaries, and their immune systems were at least as naive to European infections. To explorers and settlers, they gave the impression of being Adams and Eves in a salubrious Eden. Columbus was the first of many to admire their fine physiques, vigor, and apparent freedom from disease. Actually, the cultures and health of many Amerindians were already declining when Europeans arrived. And where newcomers saw unexplored bounty, many natives saw a fatigued land that had failed them during a long, sometimes unhappy history.

The Americas had no human inhabitants until nomadic bands followed mammoths and other big game from Siberia to Alaska over a land bridge called Beringia. That strip, about 100 miles wide, has appeared periodically where the Bering Strait now lies, when glaciers lock up much of the world's water and lower sea levels. Until recently, it was thought that migrants first crossed Beringia 12,000 years ago. Now some scientists believe Asians reached the New World 30,000 or even 60,000 years ago, some perhaps by island-hopping across the Pacific to South America. There may have been several migrations by different routes, from several sources. The Beringia crossing remains the favored theory, though, partly because it helps explain Amerindians' freedom from many Old World infections. Trekking across Beringia and then south through ice and tundra, Paleolithic nomads traversed a cold filter that eliminated many microbes, to say nothing of frailer humans.

Asian migrants scattered over the New World and developed a wide variety of languages and cultures. In some places, the hunting-gathering life gave way to primitive agriculture, then to large-scale farming and urban centers. This change began in Peru as many as 7,000 years ago; in parts of the southeastern United States, it was complete only 1,000 years ago. The staple crops were such New World plants as the potato, cassava, and maize. In several regions, vast fields of maize and beans supported cities with tens or hundreds of thousands of inhabitants, as big as any in medieval Europe. These cities, with their grand temples and pyramids, were linked by roads into far-flung empires and confederations—the Incas in Andean South America, the Mayas in tropical Mesoamerica, the Aztecs farther north in Mexico, the Anasazi in New Mexico, and the Mound Builders in the eastern United States.

These New World empires were roughly on a cultural level with ancient Sumeria, but they lacked some key technology of their Old World counterparts. Amerindians invented the wheel, but used it only for toys. They forged metals, but made them into ornaments, rarely tools and weapons. They domesticated no large animals for meat, transport, or power (except, in South America, the llama). Their lack of herds saved them from acquiring a heavy load of zoonoses and crowd diseases; it also deprived them of varied diets, and of protein reserves when crops failed.

Settled agricultural life brought some of the same problems as in the Old World, especially diseases invited by malnutrition, crowding, overwork, and poor sanitation. When Amerindians cleared forests and jungles to make room for fields and villages, they caught new diseases from wild animals and birds, some transmitted by insects. As villages became cities, parasitic and diarrheal infections resulted from increasingly contaminated water, food, fields, and dwellings. The development of social classes gave rise to well-fed rulers and priests and to sickly, ill-fed masses. Mound Builder burials show that people in high-status graves died at an average age of thirty to forty; ordinary workers died a decade younger.

Many of the new infections Amerindians acquired may never be identified. One, we know, was Chagas' disease, a severe infection caused by a protozoon related to the one responsible for African sleeping sickness. It is transmitted by *Triatoma infestans,* also called the kissing or cone-nose bug. This insect adapted some 2,000 years ago to living in houses, and it makes Chagas' disease endemic in much of South America. Some historians believe that Charles Darwin's decades of poor health resulted from the kiss of *T. infestans* during his voyage on the *Beagle.* It is no less likely than some other speculations about his chronic health problems.

All over the world, the Neolithic Revolution brought dependence on staple foods higher in calories than in proteins, vitamins, and minerals. Predictably, nutrition continued to worsen in much of the New World as intensive maize farming depleted the soil. Like Neolithic farmers everywhere, the Incas, Mayas, and Mound Builders became smaller, more vulnerable to disease, and shorter-lived than their hunter-gatherer ancestors. Because of diets rich in carbohydrates, many had dental caries and abscesses. Some suffered deficiency diseases such as scurvy. Their skeletons show stress fractures and arthritis caused by unremitting field labor. As empires grew, and tribes resisted or rebelled, warfare brought combat traumas. The rigors of hunting and of coastal and riverine fishing produced high rates of mastoid infection, often severe enough to result in deafness.

Mayan history offers a sad example of the trials of Neolithic life in the New World. The Mayan empire, with its splendid stone cities, covered some 200,000 square miles in Guatemala and Mexico, and it lasted 1,500 years. Then, a thousand years ago, the empire collapsed; the jungle reclaimed its ruins, and the people reverted to village life. Studies of Mayan bones and of the Mayas' living descendants show that they were primed for disaster by their use of the environment and by their parasites.

For a thousand years before their empire fell, the Mayas' dependence on maize and beans reduced their size and robustness. The skulls of ancient Mayan children show pitting of the orbital bones and spongy degeneration of the cranium—classic signs of acute iron-deficiency anemia. Such anemia can have a number of sources, including lack of iron intake, loss of iron in sweat or blood, or anything that impedes iron absorption, iron metabolism, or hemoglobin formation. The Mayas were subjected to all of these.

The soil the Mayas farmed was poor in iron. As a result, so were their crops, and so was mothers' milk. Their diet was also low in vitamin C, which the body needs to absorb and use iron. It lacked the proteins needed for hemoglobin synthesis. The custom of soaking maize in water destroyed most of the folic acid and vitamin B_{12} needed for developing red blood cells. Maize contains iron, but it also has phytic acid, which inhibits iron absorption in the intestine. The Mayan practice of stone-grinding maize may have altered it chemically so that iron absorption was further inhibited. Heavy sweating, unavoidable in the Mesoamerican tropics, caused more iron loss, as did intestinal bleeding caused by hookworm, tapeworm, and other parasites common in Mayan farmers. In addition, the Mayan diet lacked zinc and other substances needed for growth and for resistance to infections.

Historians have guessed that the Mayan empire collapsed because of war, revolution, epidemics, or the population's outstripping its food supply. Any or all of these may have happened; whatever the final blow, the Mayas were already weakened by malnutrition, anemia, parasites, and bacterial diseases. These plague their descendants in poor farming villages today, and similar afflictions dogged other Amerindians. Bones from the Dickson Mound burials, in central Illinois, show a fourfold increase in anemia and a high rate of bacterial infections after the beginning of agriculture.

When native peoples first met Europeans, their societies ranged from hunting-gathering to advanced Neolithic, their health from good to adequate to wretched. Precontact skeletons, mummies, and coprolites indicate high rates of dysentery, intestinal parasitic diseases, wound infection, and nonpulmonary tuberculosis (probably caught from wild birds or ungulates and then passed from person to person). They also had pinta, a nonvenereal skin disease related to syphilis, and perhaps some form of syphilis as well. They probably suffered bacterial and viral pneumonias and hepatitis A. Some historians claim, but more deny, that they had one of the milder forms of malaria. They suffered only two diseases unknown in the Old World, Chagas' disease and a bacterial infection called Carrión's disease, or bartonellosis. Most important for their future, they did not have deadly tertian malaria, yellow fever, dengue, smallpox, measles, diphtheria, typhoid, scarlet fever, or influenza. Those would all arrive with Europeans and Africans.

Europeans impressed by New World youth and vigor could not appreciate the cost of such appearances. Many native infants and children died of deficiency diseases and infections. In some tribes, infanticide culled out the less fit and reduced the number of hungry mouths. Many adolescents and young adults perished in hunting accidents, tribal warfare, childbirth, or infective complications of any of these three. Life expectancies were short. Of people whose remains were found in the Sonoma region of California, dating from A.D. 500, almost 40 percent died before age twenty. The oldest among them did not reach fifty. No wonder the natives seemed young and hardy; few but the young and hardy were alive.

Europeans perceived a land that was vast and abundant. Some of it was, but primeval forest was already rare in much of the northeastern United States. Indigenes there had been burning off first growth for hundreds or thousands of years, to aid planting and hunting. In the realms of the Anasazi and Mayas, much soil had been exhausted and abandoned. Time had stood still in Yucatán and Massachusetts no more than it had in Portugal and England. Empires had risen and fallen, and some of the people Europeans saw as simple hunters or villagers were remnants of vanished grandeur.

However robust or debilitated, Amerindians had spent millennia adapting to their microbial environment. The Americas were as untouched by Old World crowd diseases as any remote Pacific island.

Therefore one person with even a mild case of smallpox, measles, or mumps could set off an epidemic that would destroy entire cities. That is what happened, over and over again.

The worst such killer was smallpox. It was probably introduced to Europe in Roman times, and perhaps reintroduced by returning Crusaders. By 1500, it had become endemic, chiefly a children's disease to which most adults were immune; epidemics would strike new crops of susceptible children every five to fifteen years. For a while in Europe, the disease had shown diminishing virulence, but when the virus reached the New World with the Spaniards, it acted as epidemics had when they killed Romans by the hundreds of thousands and drove Huns from the walls of Rome.

Some scattered outbreaks of smallpox occurred in the first quarter-century of Spanish presence in the West Indies. Then a major epidemic arrived from Spanish ports in 1518. It killed one-third to one-half of the Arawaks on Hispaniola and spread to Puerto Rico and Cuba. That year, Cortés sailed from Cuba for Mexico; he narrowly escaped taking the disease with him. He landed in Yucatán with 800 men and set out for the Aztec capital, Tenochtitlán (later called Mexico City). When he arrived, he saw a magnificent city of some 300,000 people, built on an island in Lake Texcoco and laced with canals.

At first Cortés was welcomed by the emperor, Montezuma, who thought him a god whose coming had been predicted by Aztec legend. Later, when relations turned hostile, Cortés had to leave the city. Meanwhile, reinforcements arrived from smallpox-stricken Cuba. Among them was an African slave with a mild case of the disease. While the Spanish approached Tenochtitlán again, smallpox spread among the natives, first outside the city and then within. In 1521, Cortés attacked with 300 Spaniards and some native allies. Three months later, when Tenochtitlán fell, Cortés learned that half of its people had died, including Montezuma and his successor. The canals were choked with corpses. "A man could not set his foot down," said Cortés, "unless on the corpse of an Indian."

The disease took its typically virulent course in a virgin population. Victims broke out from head to foot in pustules so awful that their flesh came off when they moved. Many who survived were left pockmarked or blind. It was not only the great city that suffered, wrote the friar Toribio Motolinia. In most Aztec provinces, more than half the population died . . . in heaps, like bedbugs. Many others died of starvation, because, as they were all taken sick at once, they could not care for each other, nor was there anyone to give them bread or anything else. In many places it happened that everyone in a house died, and, as it was impossible to bury the great number of dead, they pulled down the houses over them in order to check the stench that rose from the dead bodies, so that their homes became their tombs.

The surviving Aztecs were stunned and apathetic, and awed by the white men who went untouched among the dead and dying. Native social and political structures shattered. The Spanish conquest, tribal and civil wars, and terror of smallpox sent people fleeing. They went on foot and by boat, along trails, rivers, and coasts. Because smallpox incubates for ten to fourteen days, an apparently healthy refugee could carry it hundreds of miles before showing symptoms. By 1530, smallpox had rushed ahead of the conquistadors, covering the Americas from the pampas to the Great Lakes.

South of the Aztec empire lay the realm of the Mayas. Smallpox reached them before the Spaniards did. It may have come with fleeing Mayas or up from Panama, where the Spanish had introduced it in 1514. The Mayas called the disease *nokakil,* the great fire, and it ran through them with searing devastation. It continued south toward the Inca empire, in Peru. Francisco Pizarro set out from Panama to find the Inca capital, but *nokakil* outran him, killing hundreds of thousands. The chief Inca died; so did his son and heir, and many nobles and generals. Years of recurring epidemics and civil war followed. In 1533, when Pizarro finally entered Cuzco to plunder it of gold treasure, the Incas were incapable of serious resistance.

A lethal crowd disease preceded the Spaniards as they ventured northward to the Mississippi valley. In 1539–1542, when Hernando de Soto made his way through the land of the Mound Builders, he found uninhabited towns where corpses were stacked in large houses. Some pestilence, perhaps smallpox ignited by shipwrecked Europeans or by Indians from the south, had done the fighting for him.

Measles followed smallpox to the New World. Both were spread by troops, sailors, missionaries, colonists, messengers, and fleeing natives. In 1529, a measles epidemic in Cuba killed two-thirds of the natives who had survived smallpox. Two years later it had killed half the people of Honduras, ravaged Mexico, spread through Central America, and attacked the Incas. The toll of smallpox and measles was ten thousand here, a hundred thousand there, whole cities and tribes wiped out, cultures and languages lost. Corpses lay scattered in fields and heaped in silent villages. Through the sixteenth century and beyond, these infections were reintroduced by Spaniards, Portuguese, French, English, and Africans. The "great fire" kept raking native peoples from Canada to Chile.

Other Old World diseases followed—mumps, typhoid, typhus, influenza, diphtheria, and scarlet fever. Some epidemics remain unidentified today. One such infection struck Peru in 1546, killing countless indigenes and even their flocks of llamas and sheep. The Aztecs, too, suffered the raging fever and nasal hemorrhages of the disease, which they called *matlazahuatl.* The word implies skin lesions other than those of smallpox and measles. Some historians suggest it was typhus, but others disagree. More than forty years later, in 1589, another epidemic struck Peru. This disease began with headache and kidney pain; a few days later, victims became stupefied, then delirious, and many ran naked and shouting in the streets. The skin lesions were so destructive that movement could make chunks of flesh fall away. The disease might strip the skin from faces and destroy noses and lips, leaving the facial bones bare. When this horror spread to Chile, it killed three-quarters of the Araucanians.

However daring and resourceful Cortés and Pizarro were, it beggars imagination that each defeated an empire of millions with mere hundreds of soldiers. Their strongest ally was the Fourth Horseman, in the form of Old World epidemics. Since they and their countrymen remained untouched by plagues that killed native peasants and emperors alike, both sides believed the epidemics were punishment by the white men's god. Natives not killed by disease were demoralized by military defeat, reprisals, forced labor, and slavery. Parents allowed babies to die, and suicide became commonplace. This misery continued for centuries.

Historians' estimates of the toll in Latin America vary, but all agree that it was staggering. In the Bahamas, where Columbus first landed, the Lucayans who escaped death by infection were sent into slavery in Hispaniola; by 1513, not one remained there. A governor of Santo Domingo said that by 1548 the island's 1 million natives had been reduced to 500. He concluded that "God repented having made such ugly, vile, and sinful people" and that "it was His will that they should die." In 1568, less than fifty years after Cortés arrived in Mexico, its total population, perhaps originally 25 to 30 million, was a mere 3 million. According to one estimate, smallpox alone killed 18 million people in Mexico in the sixteenth century. Spanish census figures from Peru show that in the years 1553–1791, the Inca population sank from 8 million to 1 million.

In North America, the natives suffered similar slaughter by European microbes. French and English settlers brought the same crowd diseases as did the Spanish. An epidemic, perhaps smallpox, spread from Nova Scotia to Massachusetts in 1616–1617, thinning the native population just before the Pilgrims landed. Not one European fell sick in 1630, when Increase Mather recorded that "the Indians began to be quarrelsome . . . but God ended the controversy by sending the smallpox among the Indians of Saugast, who before that time were exceedingly numerous. Whole towns of them were swept away, in some not so much as one soul escaping the destruction." In 1634, smallpox scoured New England again and spread west to the Great Lakes, destroying entire Algonquin towns. That year William Bradford, the governor of Plymouth Colony, described the epidemic:

> This spring, also, those Indians that lived about their trading house there fell sick of the small pox, and died most miserably; for a sorer disease cannot befall them; they fear it more than the plague; for usually they that have this disease have them in abundance, and for wants of bedding and lining and other helps they fall into a lamentable condition as they lie on their hard mats; the pox breaking the mattering, and running into one another, their skin cleaving (by reason thereof) to the mats they lie on; when they turn them, a whole side will flea off at once . . . and they will be all of a gore blood, most fearful to behold; and they being very sore, what with cold and other distempers, they die like rotten sheep . . . they were in the end not able to help one another; no, not to make a fire, not to fetch a little water to drink, nor any to bury the dead. . . . The chief Sachem himself now died, and almost all his friends and kindred. But by the marvelous goodness and providence of God not one of the English was so much as sick, or in the least measure tainted with this disease.

Thus, said John Winthrop, first governor of Massachusetts Bay Colony, "the Lord hath cleared our title to what we possess."

As in Latin America, smallpox and measles returned again and again into the nineteenth century, never absent for more than two or three decades. Amerindians perished by the village, town, and tribe. Epidemics reduced the number of Indian towns in the lower Mississippi valley by 80 percent from 1550 to 1600; thus ended the Mound Builders' civilization. In 1645, smallpox killed half the Hurons and sent the survivors in flight from the warring Iroquois, who in turn were halved by the disease in 1684. In 1738, smallpox killed half of the Cherokees in the Charleston area; in the early nineteenth century, it destroyed two-thirds of the Omahas, and in 1837–1838 almost all of the Mandans. As in Latin America, it often preceded white explorers; when George Vancouver entered Puget Sound in 1792, he was met by Indians with pockmarked faces, and found a beach strewn with human bones and skulls.

Historians' estimates of this carnage are often in direct proportion to their view of Europeans as brutally rapacious. The fact is that some whites deplored natives' suffering, while others rejoiced in it; and regardless of intentions, none of them could have stopped these epidemics. They had no better grasp of contagion and immunity than did the genial Caribs or the warlike Mayas. Contact between previously separated populations was the chief culprit in Amerindians' appalling fate. Because Old World microbes and vectors entered global interchanges, the New World population of perhaps 100 million was reduced by about 90 percent. It was a bigger disaster than the Black Death, but spread over four centuries. The native population did not bottom out and start to recover until a century ago.

The die-off eased the way for Europeans to impose their languages, religions, and political power. It also allowed them to Europeanize the landscape, by razing forests and introducing such Old World plants and animals as chickens, horses, sparrows, black rats, sugar, wheat, dandelions, tumbleweed, and "Kentucky" bluegrass. In this rich and rapidly changing landscape, what they lacked most was plentiful labor.

For centuries the European colonies were short of hands to work their farms, plantations, and mines. At first the Iberians relied on enslaved natives; the English and French, to the north, tried using European dissenters, convicts, and indentured workers. Both practices failed. The natives died of epidemics and overwork. The inflow of white laborers remained too small, and many who did come were killed by malnutrition and infections. The solution that emerged in much of the New World was importing slaves from Africa. That brought another wave of new diseases to the Americas, afflicting both natives and whites.

Europeans had already tried to create plantations in western Africa, but the diseases there were as deadly to them as smallpox was to Amerindians. Soldiers, settlers, and missionaries died in droves, in what became known as "the white man's grave." Only in the early twentieth century would large numbers of Europeans survive in tropical Africa. Until then, they had to be content with ports and trading posts there, from which they brought African workers to the neo-Europes they were creating in colonies with temperate climates.

Imported slaves escaped decimation by smallpox; the disease had ancient endemic centers in Africa, so many adults were immune. But on crowded transport ships they died of scurvy, dysentery, typhoid, and typhus. Many already were suffering from malaria, dengue, sleeping sickness, and a veritable zoo of tropical helminths. After they arrived, they were prey to additional diseases, especially respiratory and intestinal ones, through exposure to Europeans and Native Americans. The slave ships also bore two deadly infections that would reshape the New World's disease pool, yellow fever and malignant tertian malaria.

Yellow fever broke out in Cuba and Yucatán in 1648. The Old World mosquito that carries the virus, *Aedes aegypti,* flourishes in such manmade containers as water casks, so the germ and its vector easily survived Atlantic crossings. The virus found plentiful hosts in New World monkeys. Yellow fever flourished wherever its vector could, where the temperature remained above 72 degrees Fahrenheit; it also struck briefly but lethally in warm seasons in temperate ports.

In parts of Africa where yellow fever was endemic, most people had mild, symptomless infections in early childhood and remained immune for life. In the New World, when the virus struck whites, Amerindians, and American-born blacks, many sickened and died from yellow jack (the nickname taken from the yellow flag that signaled the fever's presence on ships). Cotton Mather described a yellow fever epidemic in Boston in 1693, ignited by a ship from Barbados. It was "a most pestilential fever . . . which in less than a week's time carried off my neighbors, with very direful symptoms, of turning yellow, vomiting and bleeding in every way and so dying." Yellow fever killed more than 5,000 people in Philadelphia a century later, and physician Benjamin Rush described the terror of contagion it created. "The old custom of shaking hands fell into such general disuse, that many were affronted even by the offer of a hand." Epidemics paralyzed New York and Charleston.

Malaria had even greater effects on New World health and development. It is ironic that in the early 1600s, the Pilgrims canceled their original plan to settle in Guiana, in South America, to avoid such tropical scourges as malaria. Quartan malaria was then near its height in Europe, and European settlers brought it to the Americas on their ships. Slave vessels from Africa added deadlier tertian malaria to tropical America. Deforestation and such "wet" agriculture as rice farming offered the disease ideal breeding grounds. By 1750, both forms of the disease were common; malaria ranged from the Latin American tropics through the Mississippi valley to New England. Many enslaved Africans were protected from malaria by the sickle-cell trait and by childhood exposure to the plasmodium parasite. Their American descendants, though, were at higher risk; many also suffered quartan malaria and yellow fever, just as whites and Amerindians did.

During almost four centuries of forced migration, tens of millions of Africans died of mistreatment and sickness. Fifteen million survived as slaves. In the process, they went through a brutal biological selection for endurance and resistance to disease. This winnowing made them a more valuable commodity. In the bookkeeping of a colonial landowner, a black African was worth three times as much as an indentured European worker, and more again than an Amerindian. This was the prize for having survived with greater resistance to infections.

The calamities of natives and blacks often overshadow the fact that most whites' lives were also, by modern standards, quite harsh and perilous. In the first winter at Jamestown, half the settlers died of malnutrition and disease. These rigors were more the rule than the exception. Through the seventeenth century, up to 20 percent of white newcomers to Virginia died during the first year there, chiefly of dysentery, typhoid, malaria, and poor nutrition. Those who survived and were immune to the New World's increasingly distinctive strains of microbes became known as "seasoned hands."

As whites' prosperity grew, their isolation from European crowd diseases became a disadvantage. During childhood, their immune systems remained unacquainted with infections taken for granted in the Old World's teeming port cities. Most Americans lived on farms and in villages; until well into the eighteenth century, few American cities were big enough to support many crowd diseases. These infections were brought to the Americas again and again as epidemics, afflicting adolescents and adults. Germs that did become permanently established in their new American environment might develop into local strains with characteristics different from those of their European or African ancestors.

As a result, well-off young Americans who went to Oxford or Cambridge for education risked disfigurement or death from smallpox. In fact, fear of smallpox was a major spur to the creation of American universities. Students who did survive their English sojourns faced another challenge; they found on returning home that they had to undergo several years of readaptation to New World microbes.

Similar problems confronted freed blacks in the early nineteenth century when they were repatriated to Africa. In the home of their ancestors, they were attacked by one unfamiliar infection after another. Their African genes gave limited resistance to malaria, but they lacked antibodies created during an African childhood. They were easy prey to yellow fever, tropical helminths, and the protozoon causing sleeping sickness. They died almost as easily as Europeans.

From 1500 to 1800, natives, whites, and blacks in the New World all suffered one another's diseases; a common disease pool and more or less common immunities slowly developed, but at a very high cost. Smallpox and measles kept recurring, with complications that killed children or left them blind or brain-damaged. "Throat distemper"— scarlet fever and diphtheria—killed many children, and sometimes adults. There were periodic epidemics of whooping cough, typhoid, typhus, and flu. Ships from Europe that stopped in the Indies for provisions continued to pick up yellow fever there before traveling to northern ports and setting off epidemics. Major outbreaks of yellow jack would hit New Orleans, Memphis, and other southern U.S. ports into the early twentieth century.

Malaria, both epidemic and endemic, also remained a severe problem in the United States until the turn of this century. It afflicted presidents from Washington to Lincoln, sickened hundreds of thousands of Civil War soldiers, reached California with Gold Rush immigrants in 1849, and killed uncounted Amerindians in the West. For decades after the Civil War, it was a notorious drain on the physical and economic health of the rural South. To this day, malaria, like yellow fever, remains endemic in the tropical New World.

Only in the course of the nineteenth century did the ravages of infectious disease decrease, as urban populations grew large enough to support all crowd diseases as endemics. Then, of course, they became less deadly, and occurred mainly in childhood. There were some odd twists of misfortune along the way. Smallpox became more virulent early in the seventeenth century and remained so for some 200 years. It struck a rising proportion of adults and sometimes killed as many as 40 percent of its victims. One theory holds that the virus mutated to a deadlier form as it traveled back and forth among Europe, the Americas, and Africa. Another possible explanation lies in the fact that the virus has three varieties, with symptoms ranging from mild to moderate to lethal. Whatever the reason, by 1500, smallpox, well on the way to becoming a relatively "civilized" disease, again took many lives at every level of Western society. It almost killed young Queen Elizabeth in 1562; it did kill Mary II of England, Peter II of Russia, and Louis XV of France. In 1742, almost two centuries after the Spanish-borne disease had killed the last Inca ruler, it did the same to young King Luis I of Spain.

As the New World was reaching a new environmental and microbial balance, Europeans were busy exploring the rest of the globe. They did so for the same mixed motives, from solicitous to savage, that had fueled earlier explorations—to save souls, plunder, plant, enslave, educate, heal, steal, and create dumping grounds for criminals and dissenters. The infectious diseases, including tuberculosis and syphilis, that were widespread in Europe's port cities traveled with ships' crews and settlers. The epidemic disasters of the Americas were repeated in Siberia, Oceania, and Australia.

By the late sixteenth century, Europeans were crossing the Urals into Siberia. By 1700, they were a majority there. They carried diseases such as smallpox and flu; the syphilis they passed to native women killed many and left others infertile. Voluntary migration by Europeans continued at an accelerating pace for 200 years. In the twentieth century, not only political prisoners but entire ethnic populations from other parts of the Russian empire were forcibly settled in Siberia. As in the Americas, native peoples, cultures, and languages were weakened or wiped out, chiefly by disease.

The same process took place on islands across the Pacific. In the eighteenth century, James Cook and his men carried tuberculosis to Polynesia, where it galloped fatally through native populations. Thanks to the Polynesian custom of sexual hospitality, so did syphilis and other sexually transmitted diseases. Measles killed more than 20 percent of the natives of Hawaii in 1853, and the same percentage of the natives of Fiji in 1874. In the mid–nineteenth century, the Maoris of New Zealand made their acquaintance with smallpox, measles, whooping cough, and flu; many survivors of the epidemics died of malnutrition, alcoholism, and suicide. In twenty years, between 1840 and 1860, their numbers fell from more than 100,000 to 40,000.

When whites arrived, Australia had been populated for 40,000 years or more, by the people now called aborigines. Isolated from the rest of the world, they were untouched by crowd diseases; they had mostly endemic infections of the chronic, heirloom variety. After European settlement began, late in the eighteenth century, the aborigines were ravaged for fifty years by smallpox, cholera, typhus, and flu. Then tuberculosis and leprosy arrived and took their toll. European crowd diseases were introduced repeatedly as epidemics; in this predominantly agricultural continent, cities were not large enough to sustain them all as endemics until the 1930s.

In the centuries after 1500, exploration, technology, and disease gave Europe control over much of the world. By about 1700, most of its own diseases had been domesticated to endemics; the traffic of unfamiliar microbes was mostly in one direction, from the Old World to the New. Most important, the ecological balance of humans and pathogens had changed permanently on a global scale. Pandemics, which began with the Black Death, were now a regular feature of human life. And the movement of hosts and pathogens all over the globe would keep increasing.

Eight

microbes reply

*Germs adapt to culture. Ectoparasites:
the fatal fig leaf. Of typhus, lice, and
Leninism. Ailing potatoes, human
misery. All poxes great and small.*

As Europe bestowed infections on the rest of the world, it acquired some new ones and carried them everywhere. Typhus and syphilis made their European debuts in the 1490s and quickly became global killers. Later cholera would escape its original home in India and become synonymous with sudden death. Such diseases emerged and flourished because changes in human culture—in war, economic life, hygiene, sex, and clothing—were forcing germs to readapt. Not least important among the human changes that invited new infections was clothing.

People usually do not think of clothes as an environment, but that is just what they are to microbes. To a bacterium, a droplet of body fluid is a sea, a hair or fingernail a continent, a piece of cloth a universe. Garments are also havens for ectoparasites, which live not within their hosts but upon them. As clothes increased from a tropical minimum to a year-round cover for most of the body, they created a home for a new, minute bestiary.

The first clothing, according to Genesis, were the fig leaves Adam and Eve sewed into aprons to cover their nakedness. Actually, the first wearers of clothes must have felt not shame but triumph. Their ingenuity allowed them to survive beyond the warm environment where humans first evolved. Thanks to garments and fire, they became the only primates to inhabit the entire world, even deserts and Arctic wastes. Part of the price was a large complement of fleas, lice, and bedbugs.

These ectoparasites have their own histories, shaped in part by human behavior. Each came to live on people in different times and ways, as people created niches for them. Their remote ancestors lived on the organic debris around animals' lairs and birds' nests. Some modified their claws and mouth parts to become blood feeders that cling to fur. Fleas probably shifted from other mammals to humans in the past 10,000 years. Since they must first develop in debris before finding hosts, they rarely infest animals without lairs, such as big mobile primates and carnivores. Fleas do thrive on dogs and pigs, which lived intimately with the first sedentary humans; in fact, many primitive dwellings differed little from pigsties. From such sources fleas must have hitched on to people in Neolithic villages. Fortunately, human fleas rarely transmit diseases; they are less a threat than a nuisance.

Bedbugs probably moved in with people some 35,000 years ago. Their ancestors were plant feeders, then parasites of cave-dwelling bats. Perhaps it was in Middle Eastern caves that they began adapting to late Neanderthalers or early *Homo sapiens.* The relationship was cemented when people devised permanent dwellings, which to bugs are quite like their accustomed caves. Though mentioned by the ancient Greeks and Romans, bedbugs spread worldwide more recently; they were first referred to in Germany in the eleventh century, in England in the sixteenth. The bedbug's tropical cousin, the cone-nose, may have entered people's houses only a few thousand years ago. It transmits Chagas' disease, still a widespread problem in South America, but otherwise the bedbug family transmits infection to people only by occasional accident.

Lice are quite another matter. They are probably heirloom parasites, inherited from our primate ancestors. One reason to think so is that lice tend to make very specialized adaptations, and these usually take a long time to develop. Lice are such discriminating feeders that many species will starve to death rather than drink the blood of unfamiliar hosts. Being so specialized, they had to readapt when hair receded from most of the human body, to last bastions on the head and pubis. One type of louse evolved that had claws modified to grip the dense, fine hairs of the head. Another variety developed to inhabit the coarse, widely separated hairs of the pubis (it can also live in the coarse, widely separated eyelashes). A third type, body lice, diverged from head lice after the invention of whole-body clothing. Despite their name, they live and lay eggs not on the body but in the dense fibers of fur, wool, and cotton.

Head, pubic, and body lice all became ubiquitous in temperate climates. In the West, they were helped by early Christian disdain for comfort and cleanliness; one species' asceticism created another's nirvana. Lice thrived in the Middle Ages and beyond, thanks to the continuing belief that bathing was an indulgence, an invitation to illness, or even a sin. The growing variety and availability of garments— especially of woolens during the Little Ice Age—may have helped lice proliferate, as did growing population density and crowded living. As late as the seventeenth century, etiquette lessons for Europe's nobility taught when and how to dispose of one's lice. If royalty had plentiful parasites, so did their subjects, who washed their bodies and clothes less often. Lice became rarer from the eighteenth century on, as washing became more frequent and effective. However, head lice and nits (louse eggs) still infest many children in American schools and nurseries, and in poor countries, body lice thrive on people who wear the same unwashed clothes every day.

Ectoparasites are worse than a mere itch when they carry endoparasites, companions that dwell within the body, such as bacteria. They may transmit germs by injecting them into the host or by leaving them on the skin, where tiny abrasions caused by scratching allow them entry. Many types of fleas, lice, and ticks carry rickettsiae, bacteria so small that, like viruses, they live within rather than among a host's cells. From their ancient home in fleas and lice, rickettsiae adapted to the rodents and other small mammals those creatures infested. It was from rodent fleas that rickettsiae took a lethal course to humans.

One type of rickettsiae, *R. typhi,* lodged in rat fleas and in rats so long ago that now they do no harm. When rat fleas accidentally meet and feed on people, they may pass on the germ; it causes endemic typhus, a bothersome but seldom deadly disease. This must have happened increasingly as village and urban life brought more and more contact between people and scavenging rodents. Since the germ could not be transmitted by humans or human lice, endemic typhus probably remained a minor, episodic problem. Only when a large number of people were crowded, clothed, lousy, and ill nourished did *R. typhi* evolve into a new species, *R. prowazeckii,* that could infect both humans and their lice, and thus be passed from person to person. The result was a fatal new disease, epidemic typhus. The germ's virulence is memorialized in its name; early in this century, Howard Ricketts and Stanislaus von Prowazek both died of typhus while seeking its cause.

R. prowazeckii can infect head and pubic lice, but it is usually spread by body lice. Because the disease takes advantage of malnutrition, dirt, overcrowding, and hunger, it has also been called prison fever, camp fever, ship fever, and famine fever. Often it struck along with dysentery, typhoid, relapsing fever, smallpox, and scurvy, and it was not clearly distinguished from all of these until the second half of the nineteenth century. Only in the twentieth century were the cause and transmission of typhus understood. By then it had influenced military, political, and social history the world over.

Some historians think typhus was the source of various recorded epidemics in Greco-Roman times, but it probably evolved only a thousand years ago, as large armies kept crisscrossing Europe and the Near East. The construction of fortified castles had made sieges bigger, longer, and more elaborate; now armies often spent extended periods in filthy, rat-infested forts and camps. Conditions there invited all the things that predispose to typhus—crowding, dirt, malnutrition, and rats. The first undisputed outbreak came late in the fifteenth century, when Spanish troops besieged Moorish Granada.

Late in 1489, Spain imported from Cyprus mercenaries who had recently fought Turks to the east. Soon after they arrived, Spanish soldiers began to sicken and die of a disease unlike any witnesses had ever seen. It began with a headache, high fever, and a body rash; the face might darken and swell; delirium followed, and then the stupor that gave the disease its name—*typhos* is Greek for "smoke" or "haze." The rash could lead to sores, then to gangrene that destroyed fingers and toes; victims literally rotted alive, giving off a hideous stench. Until the antibiotic era, many of those infected might die. The Granada outbreak killed with frightful efficiency; after a few months, Spain had lost 20,000 soldiers, 3,000 to battle and 17,000 to typhus.

The first detailed description of typhus came from the Veronese physician Girolamo Fracastoro, or Fracastorius. His book *On Contagion,* published in 1546, is the founding classic on infectious diseases. Fracastorius said this new disease had entered Europe from the east, and he was probably right. Typhus had the virulence typical of new infections, for humans and for human fleas. It probably evolved in the Near East during the Crusades and entered Europe at two points where Christian and Muslim armies met: southern Spain, in the 1490s, and the Balkans, in the sixteenth century.

From Granada, typhus spread through Spain, into France, and beyond. In 1528, when France and Spain were fighting for control of Europe, typhus scourged the French army besieging Naples; half of the 28,000 French soldiers died in a month, and the siege collapsed. As a result, Charles I of Spain held power over Italy, and over the papacy of Clement VII. Later, for fear of offending Charles, Clement would refuse a divorce to England's Henry VIII. Thus typhus indirectly helped ignite England's Reformation and the civil war that eventually followed.

When German and Italian armies battled the Turks in Hungary and the Balkans, typhus again helped shape empires, and it found another portal to Europe. In 1542, the disease killed 30,000 Christian soldiers in Hungary; four years later, it struck the Turks, breaking their siege of Belgrade. In 1566, the Holy Roman Emperor Maximilian II lost so many men to typhus in Hungary that he had to make peace with the Turks. His disbanded troops carried the disease back to western Europe. From there it traveled repeatedly to the New World, where it joined smallpox and measles in wiping out native peoples, especially in Mexico and Peru. The depradations of typhus continued in Europe during the Thirty Years War, from 1618 to 1648, and in the 1640s it ravaged both sides in England's civil war.

England had already learned that typhus could be as fearsome as plague, after a series of trials known as the Black Assizes. They were named for epidemics set off by lousy prisoners during their days in court. In 1577, one prisoner at Oxford became the angel of death for 510 people, including two judges, a sheriff, an undersheriff, six justices of the peace, most of a jury, and some hundred members of the university. The custom arose of judges' bearing nosegays, to ward off the stinking miasma thought to cause "jail fever." Such epidemics recurred for centuries. One that began with a single defendant at London's Old Bailey in 1750 killed, among others, the city's lord mayor. This spurred the construction of a windmill-driven ventilation device to lessen the stench in Newgate prison. Seven of the eleven men who built it came down with typhus.

Typhus took a high toll in the Seven Years War and the French Revolution. A major outbreak helped thwart Napoleon from creating the biggest empire in world history. Writers and artists have perpetuated a picture of his Grand Army disintegrating from cold and hunger during its retreat from Moscow. Actually, the Russian campaign had been lost before the army reached Moscow, largely because of typhus.

In the summer of 1812, Napoleon set out for Russia with more than half a million men. The weather was unusually hot and dry; water was scarce. Soon the army outran its food supply and was drinking, cooking, and washing with the little polluted water it could find. Hygiene on the road was primitive to nonexistent. As waves of men and animals crossed into Poland, dysentery and typhus broke out. Soon almost a fifth of the army was dead or too sick for duty. As the troops continued east, they were felled more by disease than by battle. In early September, when Napoleon entered Russia, he had only 130,000 men. After casualties at Borodino and more losses to typhus, 90,000 remained to continue to Moscow. In that city's smoking ruins, they were racked for a month by hunger and disease.

The army that began its retreat on October 19 was a sick, starving mob a fraction of its original size. Snow came in early November, bringing frostbite, pneumonia, and death from exposure. Soldiers gnawed at leather, ate human flesh, and froze in their own wastes. By the end of December, only 35,000 had made it back to Germany, many of them sick or dying. Napoleon's Marshal Ney wrote that "General Famine and General Winter, rather than Russian bullets, have conquered the Grand Army." Equal credit should go to General Typhus.

A similar fate awaited the army of another half-million men that Napoleon managed to raise in 1813 to resume his struggle for empire. That autumn, typhus again gripped central and eastern Europe. It killed half of the army and put an end to Napoleon's dream of world conquest.

Typhus remained a worse killer than combat as late as the Crimean War, in the 1850s. The nineteenth century also saw civilian epidemics in such crowded cities as London and Philadelphia. Then around mid-century, perhaps because of improving hygiene, typhus began to wane in Europe and North America. It killed relatively few people in the American Civil War and the Franco-Prussian War of 1870. However, it remained common in other parts of the world, and it would wreak one more terrible wartime slaughter, probably the worst of all.

At the start of World War I, typhus erupted on the eastern front; in some places it killed 70 percent of its victims. It wiped out 150,000 soldiers in Serbia, virtually removing that region from the war. Despite the squalor in the trenches on the western front, typhus hardly appeared there. Instead, more than a million men suffered trench fever, a mild rickettsial infection that today has nearly vanished. Whether its appearance somehow inhibited the spread of typhus remains a mystery. But typhus did spread through eastern Europe, accelerating with the collapse of civil order in Russia. From 1917 through 1921, it infected 20 million Russians and killed 3 million. At the height of the epidemic, Lenin declared, "Either socialism will defeat the louse or the louse will defeat socialism." The epidemic faded, and the louse lost, but it had been a close call.

When World War II arrived, it was understood that soap and insecticides could stop typhus. Soon vaccines would prevent it and antibiotics cure it. An outbreak in Naples in 1943 was halted when occupation troops used DDT to delouse the populace. Typhus did kill large numbers of people in Nazi concentration camps, and related scrub typhus caused many casualties in the Pacific theater. Still, for the first time in 400 years, a major war had been fought without typhus' becoming a major threat. In the years 1490–1920, the disease had killed far more people than armies had. Bacteriologist Hans Zinsser wrote:

> Soldiers have rarely won wars. They more often mop up after the barrage of epidemics. And typhus, with its brothers and sisters,— plague, cholera, typhoid, dysentery,—has decided more campaigns than Caesar, Hannibal, Napoleon, and the inspectors general of history. The epidemics get the blame for defeat, the generals get the credit for victory. It ought to be the other way around.

During those four centuries, famine was as regular a source of typhus as was war. In fact, one of history's worst bouts of typhus was part of Ireland's Great Hunger, known elsewhere as the Potato Famine. In the 1840s, the famine and its attendant epidemics removed almost 3 million people from Ireland by death or emigration. This was a pivotal event for Ireland, and an important one for several other nations. It is a classic example of the interplay among a changing ecosystem, human behavior, and contagious disease. The Great Hunger began with an ailment of potatoes and ended with a pestilence of humans.

Few people but farmers note, let alone fear, the ailments of vegetables. But just as microbes sicken and kill animals, they attack and destroy plants, from cabbages to elm trees. The first virus identified was the tobacco mosaic virus, which ruins tobacco leaves. Like animals, plants have chemical defenses against parasites, rather like the human immune system; that is why crops can be bred that resist various microbes. However, engineering disease-resistant crops is a recent technology. And when one species dominates an ecosystem, as in single-crop farming, an infection can easily disrupt life all along the food chain, bringing starvation, disease, and death to many species. That is what happened when potato blight set off the potato famine and "famine fever."

The cause of the blight was a fungus, *Phytophthora infestans,* native to the potato's original home, Peru. Potato farming spread to North America and, in the 1590s, to much of Europe; later the fungus followed. Blight ruined crops on North America's northeast coast in the early 1840s; it may have reached Europe already by ship. This wind-borne fungus grows so fast in warm, damp weather that it can turn a field of potatoes into a mass of stinking black slime overnight. Ireland, with its soft warm breezes, offered *P. infestans* an ideal second home, especially during the abnormally wet years of the middle and late 1840s.*

For Ireland, the timing was terrible. Improved preservation and transport of food had been reducing famines in the West for a century, but the "Hungry Forties" brought crop failures and economic crises in much of Europe. Ireland was less able than most nations to cope with

---

* *P. infestans* still recurs and ruins crops, endangering the food supply of poor nations. In 1993, fungicide-resistant strains struck fields in the United States, Mexico, Eurasia, and Africa.

the pressures. Subjugated for seven centuries, it was an impoverished land of absentee landlords, desperate tenant farmers and smallholders, colonial misrule, and rebellion. Thanks partly, perhaps, to vaccination, its population had risen to about 9 million, the most dense in all of Europe. Lacking the jobs and money economy created elsewhere by the Industrial Revolution, the Irish lived at the edge of survival.

As Ireland's population exploded, land was subdivided into smaller and smaller plots. Many people inhabited windowless mud cabins along with their pigs and garbage. Half subsisted chiefly on potatoes, with small protein supplements of milk or fish. Most of the grain and livestock they produced had to be sold to pay rent in cash. If the potato crop failed, farmers starved; so did many of their animals, which depended on potatoes as fodder. This happened often in the eighteenth and nineteenth centuries, because of such plant diseases as dry rot and leaf curl. Famine invariably followed. So did famine fever, a combination of typhus, relapsing fever, dysentery, and scurvy.

The blight struck Europe in 1845. Nowhere did it hit as long and hard as in Ireland, and nowhere did people so depend on potatoes to survive. By spring 1846, starvation was becoming widespread. Farmers who had grain sold it to pay rent; they sold their clothes and bedding to buy food. Those who could no longer pay rent were evicted; if they resisted, their homes were "tumbled" into ruins by soldiers. They took to scalpeens, refuges dug in the earth or in the rubble of tumbled buildings. Starving, in rags or naked, hunted even from shacks and scalpeens, they roved the land seeking scraps of food. People tried to survive on uncooked nettles. Wanderers carried typhus everywhere. Piles of bloody waste revealed which hovels housed victims of "bloody flux," or dysentery. Those with scurvy lost their teeth, and their legs swelled and turned black.

This was only a prelude. In August 1846, fog and torrential rain brought worse blight than in the year before. Authorities set up hospitals, soup kitchens, and workhouses, and English citizens raised charity funds. Relief efforts were overwhelmed by thousands and then millions of wretched victims of famine. The English government, in the grip of a financial crisis, gave too little too late. Crops were suffering in England as well; the country was importing American maize to feed its own poor. It had less sympathy and less succor for the Irish.

Then came the winter of 1846–1847, the worst in Europe's memory.

People died of exposure and pneumonia as well as famine fever. Without food or fuel, too weak to fetch water for drinking, cooking, or washing, the sick huddled together against the cold in hovels and ditches, their lice spreading not only typhus but relapsing fever, usually less lethal than typhus but a serious, recurrent infection. In some houses, the only living things were rats and dogs eating the corpses.

There were food riots, looting, and pathetic attempts at political independence. Emigration became a panicked flight. For the first time, travel to America continued through the winter, despite stormy weather. People died in transit by the tens of thousands, on what were dubbed coffin ships. Survivors of the crossing, many of them infected, were dumped on the docks of Liverpool, Glasgow, and North America's eastern ports. There they set off epidemics; only strict quarantines prevented mass contagion in New York and Boston. Hundreds of thousands would emigrate every year into the 1850s; they created an overseas Ireland that played an important role in the old country and the new ones, especially in the United States. No nation lost so large a proportion of its people to emigration in the nineteenth century.

By spring 1847, even prosperous farmers were ruined. Paupery, starvation, and disease reached Dublin's workers and shopkeepers; trade was at a standstill, and the sense of doom was general. The city, like the country, was wandered by feverish skeletons with sunken eyes, holding children too weak to cry. Driven by political and economic pressures, the British declared the disaster finished, but it had not even abated. In 1848 came another wave of blight. That year an abortive Irish uprising turned English public sympathy to fear and indignation; private charity funds dried up.

Then in January 1849, cholera appeared, perhaps carried by ship from Edinburgh to Belfast. It spread to workhouses, hospitals, and jails all over Ireland. In the hospitals, people lay two to a bed and died on the floors. English funds meant to relieve famine, inadequate to begin with, were diverted to cholera victims. Tens of thousands died of the infection. People were now asking to be transported to Australia as convicts rather than die at home of starvation and sickness.

Blight, famine, and epidemics would erupt on and off for decades, but never again with such unrelieved ferocity. From 1845 to 1851, Ireland's population had sunk from about 9 million to 6.5 million; roughly equal numbers had died and emigrated. The true numbers may well be higher, for uncounted people died in scalpeens and ditches, on outbound ships, and in nations to which they had fled. The tragedy would be blamed on politics, economics, and nature; it lives in Irish tradition as a defining national calamity. A biologist or demographer might say that Ireland's overpopulation had been solved by nature's cruelest cure, a die-off culminating in a lethal epidemic.

"Irish Potato Famine" is a misnomer in several ways. The blight struck in other countries, and people in England, Scotland, Belgium, Holland, and Germany also starved, died, and emigrated. Although the disaster began with famine, ten times more people may have died from typhus, cholera, and dysentery. Famine scourges mostly the poor, but this one brought epidemics that afflicted the prosperous; in fact, disease fatalities were higher among the well-off, who probably had acquired fewer immunities in youth than the poor. Finally, although famine is thought of as a natural disaster, the Great Hunger, like most twentieth-century famines, reflected poverty, poor food distribution, and poor government planning as much as actual food shortages. Corn and other foods were in Irish markets in 1847, but few people could buy them.

Some historians have claimed that the hunger and sickness resulted from deliberate neglect. There may have been English officials happy to see famine knock rebellious Ireland prostrate, yet no such influence on policy is apparent. By modern standards, nineteenth-century governments barely helped their own sick and hungry, let alone those of their colonies. Bare-knuckled laissez-faire capitalism saw welfare, even in time of catastrophe, as a social and moral corruption more harmful than what it might cure. Relief was reluctantly delivered, and often ineffectual. Today the English response to the Great Hunger seems hesitant and callous; even at the time, many found it appalling. But it is unlikely that, as some have claimed, the hunger and disease in Ireland were furthered in an act of mass murder.

What did happen was an unusual disaster—several successive years of bad weather aggravated by blight, overreliance on one-crop farming, an antiquated land system, political suspicions and hatreds, and administrative fumbling. The Great Hunger remains a classic example of how social, political, and economic forces entwine with biological ones to cause or worsen epidemics.

Typhus is not the only pestilence that rose from cultural as well as biological forces. Syphilis, which first struck Europe in the same decade as typhus, also depended on a growing, crowded populace and on global travel by humans and microbes. It, too, owed its existence partly to changes in climate, clothing, and lifestyle. Indeed, it is one of the purest examples of human behavior's shaping the evolution of germs.

Syphilis is called a sexually transmitted disease (STD), but "reproductive disease" might be a better term. Reproduction requires the transmission of genes, but not necessarily by sex; and even without sex, it can spread infection. Some invertebrates reproduce by a process similar to cloning in ovarian cells; certain germs have adapted to this, passing from generation to generation within hosts' ovaries. One such microbe is the LAC virus, the cause of LaCrosse encephalitis; it spreads by ovarian transmission in mosquitoes. There is recent evidence that *R. typhi* can infect generations of rat fleas the same way.

In higher species, copulation offers germs many ways to spread, by the contact of skin, mucous membranes, body fluids, and breath. Turtles transmit genital mites by coitus, and a protozoon like the one causing human sleeping sickness is passed on venereally by horses, causing the disease called dourine. Humans are an especially fertile field for such germs, because our eroticism is unmatched in nature. Most mammals have only short, seasonal bursts of sexual activity; perhaps only dolphins are as sexually excitable and active throughout life as humans. No other mammals couple in as many nonreproductive times and ways—orally, anally, homosexually, during pregnancy and lactation, sometimes before fertility starts and after it ends. This lavish eroticizing of human life may have evolved to create a pleasure bond between partners who must raise offspring through nature's longest, most helpless childhood. One consequence has been an explosion in the incidence and variety of STDs.

Tuberculosis, mononucleosis, and common respiratory infections sometimes travel from lover to lover. Other germs are more than erotic opportunists; they specialize in sexual transmission. Some are longtime human companions. Herpes is probably an heirloom infection, inherited in its oral form (cold sores) from primate ancestors; now it is common in genital form. Gonorrhea has made a special adaptation to humans; it has no animal reservoir and manages to thrive in tissues hostile to most germs (urine and the urinary tract are normally sterile). It may have arisen in Neolithic towns or in early cities. Symptoms suggesting gonorrhea were described several thousand years ago in Europe and Asia, but today one cannot distinguish what may have been gonorrhea, syphilis, or other STDs. Only in 1827 were syphilis and gonorrhea proven to be distinct diseases, not different stages of the same infection.

Untreated, gonorrhea can bring about many dangerous complications, including sterility. Among hunter-gatherers or in small villages, it could seriously affect the birth rate. However, the low number of sexual partners available to each person in a small community probably limited its spread in the distant past. The rise of cities and of prostitution greatly expanded each person's pool of potential partners, and thus the risk of STDs. Large mobile armies, accompanied by camp followers, further increased opportunities for the sexual transmission of germs. Predictably, one microbe specialized to take advantage of this; it was the one causing syphilis.

Syphilis appeared in Europe during the same warring, footloose era as typhus. Early in 1495, Charles VIII of France besieged Naples with 50,000 mercenaries. The soldiers—Flemish, Gascon, Swiss, Italian, and Spanish—were attended by 800 camp followers. When Naples fell, in late February, the conquerors indulged in a long bout of pillage and debauchery; then the soldiers and prostitutes scattered all over Europe. A few months later, chronicles of the battle of Fornovo described a new venereal plague.

The disease began with genital sores, progressed to a general rash, and then to revolting abscesses and scabs all over the body. The sores became ulcers that could eat into bones and destroy the nose, lips, eyes, throat, and genitals. There were agonizing pains in the muscles and bones of the limbs. The disease could be fatal in years or even months. In an edict of August 1495, the Holy Roman Emperor Maximilian I proclaimed that nothing like this had been seen before, and that it was punishment for blasphemy. Almost everyone agreed that it was a new disease; most thought it was communicable and related to sex.

By the end of that year, syphilis had aroused terror in France, Switzerland, and Germany. By 1500, it had spread to Denmark, Sweden, Holland, England, Scotland, Hungary, Greece, Poland, and Russia. STDs, like scurvy, were virtually occupational diseases of sailors and soldiers; they carried syphilis everywhere during the age of exploration. Vasco da Gama took it to Calcutta in 1498. By 1520, it had reached the north and south of Africa, the Near East, and the coast of China. Soon it would strike Japan and run rampant in the New World, and explorers and settlers would carry it to Siberia, Australia, and Oceania. Science later learned that it could pass to fetuses in the womb, and that it often caused not only sterility but paralysis and psychosis. Everywhere syphilis raised the death rate, reduced healthy births, and further weakened epidemic-ridden colonial populations.

With its painful, repulsive symptoms, the disease replaced leprosy as a badge of sin and pollution. People around the world named it for the nations they thought had infected them; in France it was the Italian disease, while in Italy, Germany, and England it was the French disease. It was the Spanish disease in Holland, the Castilian disease in Portugal, the Polish disease in Russia, the Russian disease in Siberia, the German disease in Poland, the Christian disease in Turkey, the Turkish disease in Persia, and the Portuguese or Chinese disease in Japan. It became the most disowned infection in history. The most common name for centuries was *grande vérole*, or great pox, for the rash that appears in its early stage.

The word "syphilis," not in general use until the nineteenth century, came from the same Fracastorius who first described typhus. He was not only a great physician but an accomplished poet. In 1530, sixteen years before his book on contagions appeared, he published a long poem in Latin, *Syphilis sive morbus gallicus* (Syphilis, or the French Disease). It told of a shepherd named Syphilis who, having enraged Apollo by blaspheming, was cursed with a disgusting and dolorous new disease. This gruesome pastoral went through more than a hundred editions in the sixteenth century.

In his 1546 book, Fracastorius described syphilis in detail and set the stage for centuries of debate about its origin. In the past twenty years, Fracastorius wrote, the symptoms of syphilis had become less florid and severe, with fewer ugly pustules and less pain. Other witnesses agreed that now it caused less bone destruction and fewer deaths than before. Asian chroniclers also would say that the infection's first fury soon abated. These descriptions suggest a new disease that quickly became less virulent, as parasite and host developed more tolerance of each other.

That may have been the case, but some later writers were skeptical. Perhaps, they said, Fracastorius erred, and the disease only seemed less severe. Perhaps 1495 saw not the debut of syphilis but a flare-up of a rare form of the disease called malignant syphilis. Perhaps syphilis had existed in Europe before the 1490s and had been confused with other infections. The origin and nature of syphilis became one of medical history's most intriguing disputes.

Virtually every witness to the early outbreak maintained that syphilis was a new disease, or at least new to Europe. Then where had it come from? For centuries, people repeated Fracastorius' answer, that Spaniards had brought it back from the Indies. Maybe some of Columbus' sailors, after returning in 1493, joined the mercenaries marching with Charles VIII on Naples, where the disease first broke out. If that was true, syphilis was the only major disease to travel from the New World to the Old during the age of exploration. That has struck some people as meager justice, given the heavy traffic in the other direction.

The evidence has never been fully convincing. The idea appeared after a suspicious delay; it first saw print in 1518, more than two decades after the outbreak. It was later supported by the former colonial official Oviedo y Valdes, who said he had seen syphilis in New World natives, and by Diaz de Isla, a Barcelona doctor who claimed to have treated Columbus' crew for it. Fracastorius was repeating these belated revelations a half-century after the fact. It is impossible to detail here the doubts that have been raised. Bacteriologist Theodor Rosebury said fairly that defenses of the New World theory often have an odd quality "faintly suggesting an effort to get through to the totally deaf by shouting."

Many historians deny that syphilis came from the Americas. Rather, they say, it existed in Europe long before Columbus, probably smoldering in mild form. It may have been lumped with other disfiguring diseases under the term "leprosy" and become distinctively visible after the Black Death, when leprosy faded. Or perhaps mutations produced a new strain of an old germ, giving it greater virulence. There is also an argument for the disease's having African origins.

One would expect the question to be settled by the evidence of syphilitic bones from pre-Columbian times, in either the New or the Old World. Unfortunately for scholars, though not for patients, only a small minority of syphilis victims suffer lasting bone damage, and it cannot always be distinguished from the marks of other diseases. There is little argument that bones marked by syphilis or something very much like it have been unearthed in the New World. Similar discoveries in Europe have been few; the age and diagnosis of almost all are disputed. Two recent finds, from Italy and England, are now under study and debate. Meanwhile, opposing scholars brandish conflicting theories and data at each other, making few converts.

A pair of theories have emerged that put syphilis in a broad ecological and evolutionary perspective. Although they differ in some details, they both hold that syphilis and three other conditions—pinta, yaws, and bejel—reflect microbes' response to changing human culture. All four diseases are caused by a delicate, corkscrew-shaped bacterium called a spirochete or treponeme. When the germ was isolated, in 1905, it was named *Treponema pallidum,* or pale thread, because it showed up white under a dark-field microscope. It was indistinguishable by any test from the germs causing pinta, yaws, and bejel. The question remains whether these are four closely related species, four strains of one species, or one germ whose effects vary with climate and means of transmission. C. J. Hackett believes that all four arose from the treponeme's mutations in response to a shifting environment. Ellis Hudson thinks the four diseases are actually one, which takes different forms with changes in transmission and human lifestyle. What follows conflates their theories, though leaning a bit more on Hudson's.

The treponeme's common ancestor first probably lived on decaying matter and then became a nonvenereal parasite of African primates. About 20,000 years ago, it created a zoonosis in humans. This was pinta, a rather mild but disfiguring skin infection of children and young adults. It spread through much of the Old World, crossed Beringia with ancient migrations, and today is limited to tropical villages of Latin America. In that climate, children's bare, perspiring skin transmits the spirochete through casual body contact. Some 10,000 years ago, a mutation in the germ gave rise to yaws, probably in Africa. Like pinta, yaws usually affects the skin of the young; however, it is more severe and can erode the bones. It persists today in the rural tropics of Africa and Latin America.

A few thousand years later, the germ spread to Neolithic villages in dry, cool environments, where people went fully clothed. Because clothes interfered with the germ's passage to new hosts, it retreated to the warm, moist refuges of the mouth and, secondarily, to the genitals. Transmitted chiefly by common eating utensils and sometimes by kissing, it led to a new disease variously called bejel, endemic syphilis, or nonvenereal syphilis. More severe than pinta and yaws, bejel can damage the bones and the heart. It once flourished in European slums from

Russia to Scotland (where it was known as sibbens), but it faded as hygiene improved. It is still common in villages in arid and semi-arid parts of Africa and Asia.

Venereal syphilis emerged some 6,000 years ago, in the Middle East, once the bejel germ had adapted to urban life there. More sexual partners were available to everyone, and coitus became the usual means of transmission. Venereal syphilis was not limited by climatic conditions, and eventually it spread worldwide. Opportunities for transmission were less frequent than for pinta, yaws, or even bejel; the germ survived by lingering in the body for long periods, wreaking slow havoc on the heart, nervous system, and other organs. Thus treponemal sickness was transformed from a mild disease of village children to a serious one of urban adults.

If this theory is true, treponeme infections took a course unlike that of most zoonoses, which usually shift from acute diseases of adults to milder ones of children. However, a germ often causes severe symptoms when it inhabits unaccustomed body tissues, and there is much evidence for the Hudson–Hackett view. Only a single kind of treponeme infection is common in any region; each gives immunity to the others. This confirms that the germs are, if not identical, very closely related. Furthermore, one treponemal disease can replace another as conditions change. Syphilis has ousted yaws in Venezuela, New Guinea, and parts of Africa as people moved from villages to cities. And when people with yaws move from tropical lowlands to cool mountain areas, they lose the sores of yaws and develop bejel. Bejel and venereal syphilis have each been reported to change into the other. Such shifts have probably happened many times in the past and may still be occurring.

The germ's tendency to change in transmission, symptoms, and virulence makes the complexity of its disease manifestations less puzzling. It also suggests an alternative explanation for the 1495 outbreak in Europe. When syphilis first struck there, its florid symptoms resembled acute yaws almost as much as syphilis. Perhaps yaws was brought to Europe from tropical Africa by explorers and traders in the late fifteenth century. Aided by mutations, it may have adapted to the temperate climate and urban conditions, becoming a forerunner of venereal syphilis as we know it, with the virulence of a new disease. There may even have been more than one introduction of yaws; the slave trade probably took it from Africa to Haiti, and from there to

Europe. Thus European syphilis could have resulted from introductions of tropical yaws from multiple sources.

The debate on syphilis has broadened rather than simplified; it remains unsettled. Some say pinta was the original treponeme infection; others say it was yaws. Syphilis may have arisen, as Fracastorius claimed, in the New World, evolving there from pinta or yaws, and reached Europe by ship in Columbus' time. Hackett thinks syphilis or some precursor infection reached Mediterranean Europe from Africa by Roman times. Hudson believes treponeme infection was global by the year 1000, for the most part in mild, chronic forms. Hackett and Hudson do agree that it mutated to a new, more destructive form in Europe in 1495. Regardless of which theory garners more support, the debate has focused attention on the power of culture to drive the evolution of microbes. It is an important subject still, in a time when changing lifestyles shape and reshape the course of epidemic AIDS.

Syphilis and typhus were typical new plagues of the first age of global exploration and conquest. They came from the new machinery and tactics of war; the hunger and dirt of bigger, denser populations; altered clothing and sex behavior; changing agriculture; the movements of soldiers, traders, and uprooted peasants. Early in the Industrial Revolution, similar forces would create another major new pandemic, cholera. It turned out to be the last new disease of one era and the harbinger of another. First it would spread world terror. Then it would become the first new disease humans felt they had understood and conquered.

Nine

victory, it seems

*Waters of life and death. The fright
from Calcutta. Plague brings paranoia.
Chadwick, Snow, and the pump.
Reverence for soap; health becomes public.
Laboratory heroes. From famine to health.
The forgotten flu and a husky memorial.*

Life arose in water and remains bound to it. People have always settled
on shores and riverbanks. They drink the water, wash and cook in it,
hunt for food in it, use it in crafts and industries, give it to crops and
livestock, play and cool themselves in it. Yet as human numbers grew,
people poisoned the stuff on which their lives depended. Pollution,
especially by sewage, invited new diseases that have probably killed
more people than smallpox and bubonic plague. Cholera, the most
fearsome of all waterborne infections, first appeared as Europe reached
unprecedented wealth and power. Its spread provoked global terror, and
the response to it ended up changing the length and quality of human
lives. There followed a century almost without major new infectious

diseases. The result was false hope that all infections would be conquered.

By the late eighteenth century, the Industrial Revolution was under way, and urban populations soared. If all of human history were on film, this era would seem a frantic mob scene, at high speed. Social change had never run at such a pace. The shift from nomadism to farming had taken millennia; the rise of industry and the megalopolis spanned only a couple of centuries. From the Neolithic dawn to 1820, world population rose from 5 million to almost a billion; much of the increase came at the end of that span. After population surges such as those of the Roman era and late Middle Ages, famine and plague had erased much of the gains. The growth that began in the seventeenth and eighteenth centuries was unique; it continued at an ever faster pace, and it runs out of control to this day. The new disease that signaled the dangers of this population growth and of new technology was cholera.

The growth of cities still depended largely on in-migration from farms and villages. Many of the migrants were driven by atrocious rural health conditions. Ill fed and overworked, they had lived with their animals in hovels strewn with dung and garbage. Conditions in the cities were often even worse; certainly they were different. Crowd diseases challenged newcomers' less seasoned immune systems, with dire results. Epidemics thrived on ever denser crowds. By the early nineteenth century, cities had factories, slums, and great heaps of filth; the air and water became poisonous. Railroads and ships linked urban sinkholes with faster, more frequent travel. Traders, migrants, and armies carried local infections across and between continents. Conditions were ripe for the spread of microbes between places as distant as New York and Calcutta. It was from the Ganges River delta, where Calcutta lay, that cholera sprang forth.

In 1830, rumors reached Europe of a previously unknown disease spreading out of India. Despite speculation that it might head west, alarm was muted. England had suffered no lethal epidemic since its last visit by bubonic plague, almost two centuries earlier. For a hundred years, France had seen no deadly pestilence but typhus. Europeans had realized that tertian malaria and yellow fever could visit but not take root in temperate climates. So when "Asiatic cholera" emerged from a distant corner of Britain's empire, there was an uneasy murmur but no panic. Some people doubted that cholera was a new disease; they expected it to be no worse than other, familiar ones.

Scholars debate whether cholera is, strictly speaking, new. We have seen that new diseases arise when a germ adapts to humans from their shared environment or from other hosts, and then spreads. It is not certain when the cholera germ, *Vibrio cholerae*, established itself in humans. Aidan Cockburn, an epidemiologist who studied the disease firsthand in India, concluded that it was quite new to people, a result of *V. cholerae*'s adapting to water-storage tanks and ponds in Bengali villages. Others say cholera is older but remained a local problem; it usually stayed within the Ganges delta. Perhaps it was carried through Bengal from time to time by religious pilgrims, and even now and then by ship to China's coast.

Those who argue for the antiquity of cholera point out that almost 2,500 years ago, Sanskrit writings described an affliction that resembles it; the symptoms were violent vomiting and diarrhea, a haggard face, blue lips, and muscle spasms. But if cholera did exist long before 1800, it remained limited to Bengal. Everywhere else in the nineteenth century, except part of the Chinese coast, people seemed sure that it was a new disease.

Vibrios are a family of comma-shaped bacilli whose name was inspired by their vibrating wiggle. *V. cholerae* sickens only humans, but it has many relatives in water and in marine creatures. Perhaps it evolved from a free-living vibrio in the Ganges and adapted to the human gut. Or perhaps it reached people from infected fish, or from water polluted by domestic animals. Then it was perpetuated like most diarrheal diseases, by a cycle of human infection and water pollution. People caught cholera mostly by drinking or using tainted water, but also from infected fish, vegetables washed in polluted water, contact with dirty hands or soiled linens, and flies carrying vibrios on their bodies.

Once swallowed, *V. cholerae* multiples in the intestine and releases a strong toxin. The result is vomiting and diarrhea so violent that soon the body is perilously low on fluid. Dehydration can lead to muscle spasms, shock, circulatory collapse, and death. Untreated, cholera kills from 20 to 50 percent or more of its victims, usually in days but sometimes in hours. The sick and convalescing excrete huge numbers of bacilli, which may end up back in local waters. Infected people who do not show symptoms can spread the disease as they travel. One apparently healthy visitor to a town, one flush of wastes by a boat, one meal of tainted fish or vegetables can spark an epidemic and create a permanent new home for vibrios in local waters.

In 1817, when the first cholera pandemic began, its spread was fostered by changes resulting from colonial rule. The British had founded Calcutta in the seventeenth century as an administrative center; it grew far beyond anyone's expectations, into a dirty, teeming warren. For the first time, India was laced with new roads, railroad tracks, and busy ports; there was constant movement by merchants, administrators, troops, and religious pilgrims. If a waterborne microbe could dream, that is what it would dream of. Doubtless launched by travelers from endemically infected villages, cholera began to race through Bengal's cities.

In Calcutta and Jessore, cholera killed 5,000 British soldiers within weeks. Troops carried it throughout India, then to Nepal and Afghanistan. Calcutta was a hub of world shipping; vessels took cholera from there to China, Japan, and Southeast Asia. Slave traders transported it to Arabia and East Africa. From Mecca, pilgrims dispersed it throughout the Arab world. Everywhere it struck virgin populations with swift ferocity. The first pandemic lasted six years and made cholera as feared as bubonic plague had been. By the end of the nineteenth century, cholera would sweep out of Bengal and circle the world six times. A seventh pandemic followed in the mid–twentieth century, and an eighth may have started in 1993.

When the first pandemic ended, in 1823, it had stopped short of Europe. The second began in 1826; it followed trade and troop routes to Afghanistan, Persia, and southern Russia. In 1830, Moscow became the first European city to suffer cholera, with a death rate over 50 percent. Panicked Muscovites filled the roads to St. Petersburg and Smolensk, spreading the infection. It went on to Poland and Germany, and from there by ship to England. Usually it first hit port cities, then spread along rivers and canals, and traveled by road with merchants, laborers, and refugees.

The word "cholera" entered languages and folklore all over the world. The worst curse my grandparents knew was, "May you have cholera." Growing up in late-nineteenth-century Russia, they had waited helplessly to see who in their town would die a swift, ugly death and who would be spared. It was that way throughout Europe. In the 1830s, cholera killed people by the tens of thousands in Paris, London, and Stockholm. From England it spread to Ireland; from there, immigrants took it to Mexico, Cuba, and the United States. In 1832, Mozart's librettist Lorenzo da Ponte arranged for the visit of an Italian opera troupe to cholera-stricken Manhattan. They arrived to find the streets empty and silent except for the ringing of church bells and the rattle of carts taking corpses to graveyards. Every resident who could had fled, especially northward to such rustic havens as Greenwich Village and Harlem.

In England, cholera assaulted not only bodies but some people's pride in race, class, and nation. The disease seemed a product of colonial backwardness. Its symptoms, to say nothing of its stench, were a humiliating fate for Victorian gentlefolk. They saw that cholera afflicted them less than it did the poor, ill fed, ill housed, dirty, and drunk. To chronically moralizing Victorian minds, the lower classes' weak resistance to disease proved their physical and moral inferiority. By falling prey to cholera, they showed that some of the "sceptered race" lacked industry, self-control, and godliness—sad testimony to their fitness to dominate the world's wogs. Cholera's early visitations to the United States provoked equally intense fits of moral ague; there were similar preachings and days of prayer, though with less stress on class distinctions.

Cholera also humbled the medical and civil establishments. In 1830, when accounts arrived in London that India's savage epidemic had reached Russia, doctors and government officials debated whether it was a new disease, even whether it was a distinct disease at all. Was it different from bloody flux? Did it rise from foul air or foul soil? Most important, was it contagious? A large majority of doctors insisted it was not. The idea that minuscule creatures caused disease had recurred sporadically since Democritus' day, and it enjoyed a slight vogue in the late eighteenth century. Now it was in very low repute, thanks in part to recent French experience with yellow fever.

Early in the century, France had sent 33,000 soldiers to Santo Domingo to suppress the rebellion, led by Toussaint L'Ouverture, that would bring Haitian independence. Within months, almost 90 percent of the French had died of yellow fever. The loss helped persuade Napoleon to moderate his New World ambitions and sell the vast Louisiana Territory to the United States. Some scientists maintained that yellow fever was contagious; the possibility, however slight, hung ominously over military policy-making. When yellow fever hit Barcelona in 1822, French doctors decided to study the epidemic there and settle the issue. They concluded by approving the dominant theory of the day: A miasma rising from decayed organic matter both caused and spread yellow fever. They derided the notion that invisible organisms slaughtered armies; that was like saying fleas killed elephants. No one even suggested anything as ludicrous as mere mosquitoes spreading a lethal disease. Medical opinion throughout Europe concurred. A decade later, when cholera arrived, those who believed in germs were seen as cranks, contagion theory as a relic of an ignorant age.

Despite their air of authority, doctors themselves were in low repute. Science and technology were multiplying their marvels, winning admirers and even worshippers. Medicine, to the contrary, had hardly advanced since the Middle Ages. It relied on miasma theory and on treatment by purging and bleeding. Hospitals were places were people went to die, not to be healed. Laymen unenlightened by current medical wisdom did believe cholera was contagious, and as the pandemic approached Europe, they clamored for quarantines and cordons sanitaires. Doctors felt they knew better, but they had nothing else to offer. When public apprehension verged on panic, it was impossible to do nothing, however correct that might be. Skeptically, medical and government officials revived the quarantine measures of the last bubonic plague epidemic.

Enforcing such measures was problematic from the start. No nation had yet created a public health system. England, more advanced in this matter than most nations, possessed only rudimentary vaccination and lunacy boards, and "fever hospitals" for typhus victims. Health was considered a private matter; even basic hygiene measures were widely seen as assaults on individual liberty. Throughout Britain and continental Europe, police, soldiers, and specially appointed medical officials would have to enforce the quarantines. They exercised their authority disproportionately on those whom cholera hit hardest, the poor.

As expected, cholera had some respect for class lines, but not for the moral reasons recited from pulpits and in editorials. The affluent could lock themselves in relatively clean houses or, even better, in isolated country retreats. They had more access than the poor to clean water and food, and little or no work exposure to vibrios. Cholera took an especially high toll among sailors, launderers, innkeepers, and others whose labor put them in daily contact with vibrio-laden water. Because of better personal hygiene, if cholera did strike one member of a prosperous family, it was less likely to afflict the others; among the poor, dirt and crowding enhanced secondary transmission. The vibrio was also crueler to people whose inadequate nutrition made them vulnerable to any infection.

As the epidemic worsened throughout Europe, the mood of the poor turned ugly. The quarantines they once demanded had crimped travel and trade; consequently, food prices rose. Protests and riots followed. The greater disease resistance of the rich seemed proof first of unfairness, then of oppression, and finally of conspiracy. Rumors spread that cholera was not a disease but poisoning by agents of the rich, who wanted to dispose of the troublesome poor. Such paranoia was not unique to this pandemic; it sprang up during other plagues, before and after cholera. In the Middle Ages, outbreaks of bubonic plague sparked massacres of Jews, who were accused of spreading the disease by poisoning wells. In the age of AIDS, frustrated and angry black Americans, suffering a disproportionate number of AIDS infections, have charged medicine and government with creating HIV as an instrument of racial germ warfare.

A sense of helplessness before cholera unleashed similar fury and fantasies. In Russia, cordons sanitaires provoked disorder, then violence; peasants massacred doctors and magistrates trying to enforce health edicts. In Hungary, where cholera killed more than 100,000 in the summer of 1831, peasants who thought they were being poisoned sacked castles and killed doctors, army officers, and nobles. In Prussia, tales spread that doctors were receiving three thalers from the king for every cholera death; mobs beat and killed physicians and government officials. Paris saw riots and the stoning of doctors. In some places, cholera was deemed a British fiction meant to mask the poisoning of restive Indian subjects. In England itself, riots against doctors had a different rationale. Recently the notorious Burke and Hare, not content to plunder graves for corpses to sell to anatomy schools, had resorted to gruesome serial murders. During the cholera panic, doctors were accused of using the disease as a cover for murdering patients in order to dissect them.

Cholera leaped quarantines with ease, so they were soon abandoned; there was no preventive measure to take their place. At first, few people connected cholera with urban squalor and environmental pollution, much less germs. Conditions in many European capitals matched those of Third World cities today. London's population rose 20 percent in the 1820s alone. People lived packed in tiny rooms, dripping basements, and airless attics. Coal smoke blackened the air; poor ventilation and crowding, at home and at work, aided the spread of airborne diseases. Garbage and refuse were collected irregularly or not at all. The city stank. Animal manure filled the street—and the many homes in which people kept pigs and chickens. Houses commonly relied on cesspits that were emptied into streets or open ditches. The poor saved their feces in cellars for sale as night soil. The city's seven sewer systems were uncoordinated and relied on defective pipes. They received tons of human and animal feces, dead animals, waste from abattoirs, effluvia from hospitals and tanneries, the occasional human corpse, and contaminated groundwater from cemeteries.

All of London's refuse ended up in the Thames, which provided most of the city's water. When cholera arrived, there were eight separate water companies and just one experimental filtration system. Water not taken from the Thames came from wells, many as badly polluted as the river. The city drank, cooked, and washed in its own filth. With the crowding and dirt, once a waterborne disease was established, further person-to-person transmission was virtually assured.

There were European capitals still denser and dirtier. London could sprawl outward, but Paris grew upward and inward; as a result, it was more thickly populated. With its groaning water and sewage systems, Paris lost more people to cholera than London did. Stockholm, built on a veritable sponge of low-lying water, suffered even more.

The second pandemic—England's first epidemic—faded in 1838. No one knew whether cholera would return or, if it did, how fiercely. It came back during the third pandemic, a decade later, and the toll was even higher. During cholera's absence, London's population density had risen from 35 to 50 people per acre. Therefore cholera mortality rose; having killed 2,600 people in the peak cholera months of 1832, it took almost 11,000 in the same period of 1848. The only innovation that offered hope was the nascent sanitary reform movement, yet apparently attempts to clean the city had failed in their purpose. People who opposed sanitation laws as useless, expensive, and intrusive scorned the reformers and their leader, Edwin Chadwick.

Chadwick, zealous and arrogant, embodied the era's evangelical spirit. Convinced that disease rose from dirt, he determined to wash the Great Unwashed and scour their homes and streets. In a report on poverty and health in 1842, he backed his argument with comparative statistics on dirt and disease. The numbers had been produced by physician and statistician William Farr. Farr, like Chadwick, was no slouch at preaching sanitation; he said that lethal miasmas were like a mad dog prowling forth from the city's cesspools and sewers. He was not entirely wrong.

At first, calls for improved water, sewage treatment, and ventilation aroused as much protest as support. Then cholera's second coming, so much worse than the first, renewed panic. Cholera killed fewer people then typhus, but it killed quickly and horridly, and in some places more than half its victims. So despite early failures, sanitationists got a second crack at cholera. From 1848 through 1854, Chadwick was commissioner of England's new central board of health. Laws soon made government responsible for collecting refuse, building sewers, cleaning the water and air, and tearing down slums. Cholera was called "the reformer's friend."

These measures could not stop London's second epidemic, since vibrios, not dirt, caused cholera, and the water was still full of them. In fact, some reforms made things worse; new sewers were more efficient than the old ones at flushing vibrios into the Thames. But London was terrified by cholera's return, and revolted by its own stench, so Chadwick's movement survived early failures and public resistance. It promoted government by expert committees, health legislation, and preventive medicine. Above all, it convinced people to accept that individual health and behavior were public business.

Improvements in the water system paid off in 1868, when cholera struck England a third time. There were fewer infections and fewer deaths. Although partly mistaken in both theory and practice, Chadwickians reduced cholera and other waterborne diseases, such as dysentery, shigellosis, and typhoid. People who still opposed government intrusion into personal behavior, and the tax costs of sanitation, were fighting a rearguard action.

England's success impressed governments everywhere. New York City, after three bouts with cholera, created a health board based on the English model in 1866, and the U.S. federal government soon followed. So did local and central governments throughout North America and Europe. Continuing improvements of water and waste systems reduced diarrheal diseases even though cities kept growing. In the 1890s, when cholera again ravaged many nations, Europe and North America went almost untouched. Soon chlorination, cholera vaccination, and other new preventive techniques helped eliminate the disease in the West for a century.

These later measures could come only after the cholera germ had been discovered. The link between cholera and water had been found much earlier, but it was widely ignored or disputed. A London doctor named John Snow completed decades of research with a report in 1854 that revealed one of medicine's landmark discoveries. A few years earlier, while mapping London cholera cases, Snow noticed a lethal cluster in one neighborhood. The sick and dead had one thing in common; all drew their water from a pump on Broad Street, near Soho. Snow asked local authorities to remove the pump handle. It was done, and cholera abruptly vanished from the area. Snow made more maps of cholera and water use, and traced other local outbreaks to the Thames and to polluted wells. He concluded that cholera was caused by something germlike in the water and then transmitted from person to person.

John Snow and the Broad Street pump would enter science lore along with Newton's apple and Watt's whistling kettle. His study of cholera and the statistical approach of William Farr created a model that is still used for understanding disease transmission. But as long as the miasma and sanitation theories prevailed, Snow's ideas would be deemed as outlandish as germ theory.

Germ theory and cellular biology would eventually win the day, after crucial research in the 1860s and 1870s. By the early 1880s, Louis Pasteur and Robert Koch had proven that specific germs caused specific diseases. In Paris, Pasteur found the anthrax bacillus, devised a rabies antitoxin, and began to convince the world of the principle of contagion by germs. In Berlin, in 1882, Koch isolated the tuberculosis bacillus and set forth his four postulates for proving that a microbe causes a disease. The next year, while studying the cholera epidemic in Cairo, Koch found *V. cholerae* in victims, in drinking water, and in food. This opened a new era in medicine and public health.

Now medicine began to catch up with the rest of science, in an astonishing burst of discovery and invention. For 3 million years, the average life had spanned a few decades, more or less; sudden death was ordinary at any age. Until the Neolithic, trauma and accident were the most common killers. With the coming of sedentary life, nutrition and longevity declined; famine and infection became the leading causes of death. For the next 10,000 years, it was common for microbes to strike people down at every stage from infancy to late life. Then suddenly, in the late nineteenth and early twentieth centuries, hardly a year went by without a major discovery about the cause, transmission, prevention, or cure of infectious disease.

Researchers learned the causes of most major crowd diseases, of malaria and leprosy, of childbirth fever, of infected traumas and surgical wounds. Vaccines could prevent some diseases, antitoxins cure others. Children no longer routinely risked suffocating from diphtheria or dying from the dehydration and shock of diarrheas. People bitten by animals or suffering infected puncture wounds no longer faced certain death from rabies or tetanus. With attention to antisepsis, hospitals and surgical theaters ceased to be charnel houses. By the early twentieth century, Patrick Manson, Walter Reed, and others had shown that insects and other arthropods spread malaria, yellow fever, and typhus; control or cure became possible. Paul Ehrlich created Salvarsan, a "magic bullet" for treating syphilis; it was the first chemical agent devised to fight a specific germ. As Ehrlich hoped, his work pointed the way to developing such drugs as sulfonamides and antibiotics.

Many pioneers of microbiology were brilliant and brave; no few were killed by the germs they studied. Pasteur, Koch, Reed, Ehrlich, and others were subjects of best-selling books and Hollywood movies. Even reduced to figures in medical soap operas, they shone. For several generations, young people were inspired to pursue medical science by such books as Paul de Kruif's *Microbe Hunters* and by films with Paul Muni as Pasteur and Edward G. Robinson as Ehrlich. (The tight-buttoned Koch was too severe even for Hollywood's fictioneers to romanticize.) The battle against infectious disease was rich with heroes, martyrs, rescued victims, and prodigies of detection.

For a short time, it seemed that every microbial culprit would be unmasked. There were, predictably, some false arrests; for decades a bacterium was thought to cause influenza (it is actually caused by a virus), and another bacillus was called the villain in malaria (the real one is a protozoon). There were also inadvertent comedies, as when researchers dreamed of a universal germicide to end all infection. Sir Joseph Lister had miraculously prevented surgical sepsis by dousing the operating room and everything in it with carbolic acid. What if some mild acid, swallowed or injected, could do the same inside the body? Of course, anything that killed most germs would also kill most body cells, but ambitious chemists pursued the grail of internal antisepsis. At the Bayer laboratory in Switzerland, researchers rejoiced when a mild acidic compound relieved TB patients' fever and pain. When they realized that it gave not cure but only temporary relief, they shelved it in disgust. Later the compound, acetylsalicylic acid, would be rediscovered and named aspirin.

Most of the mistakes are now forgotten or reduced to footnotes in a chronicle of triumph. The average human life span has almost doubled since the nineteenth century, and many doctors have never seen the epidemic diseases that terrorized their parents. Infant death is an exception, not a commonplace. Medicine receives most of the credit for this progress, but in reality it played a limited role. Health and longevity began improving long before Pasteur and Koch, before Snow, even before the arrival of cholera. The very population explosion that invited cholera was in part a result of improving health conditions.

Human numbers began surging as early as the eighteenth century, thanks chiefly to improved farming, food storage, and transport. Starvation, a peril as long as people depended on intensive agriculture, now became rare. The Great Hunger in nineteenth-century Ireland was especially shocking because it seemed a throwback to the bad old days of feast or famine. Demographer Thomas McKeown makes a strong case that as early as 1650–1750, improved farming was reducing deaths in the West and in China. More abundant proteins, especially from meat and dairy products, did more than hold famine at bay; they reduced childhood infections and infant mortality. As we see in Third World nations today, even small improvements in diet can reduce early-life infections, from diarrheas to common viral diseases. The effect is always to make populations soar.

Through the nineteenth century, says historian William McNeill, a race was on between ills and skills. The ills were bigger and denser populations, industrialization, and the conditions they spawned; the skills were better farming, then better hygiene, then improved medicine. Farming and hygiene were winning the race before medicine could contribute much. Despite cholera and typhus, and without antitoxins or vaccines, health and longevity improved, and death rates fell, especially for infants and children.

In the early nineteenth century, when causes of death were first regularly recorded, fatal infections were apparently already waning. Tuberculosis declined well before Koch discovered the tubercle bacillus. There was no effective drug to treat it until streptomycin became available in 1947, and no widely available vaccine until the 1950s. By then, TB had diminished steadily in the West for a century and a half, retreating to afflict the malnourished and otherwise predisposed. Diphtheria was a dreaded child killer until an antitoxin was developed in the late nineteenth century, but by then, its virulence was already lessening. Scarlet fever often killed or disabled children until sulfonamides appeared, in the 1930s; however, it had been declining in prevalence and severity for half a century or more before that.

As McKeown acknowledges, nutrition alone does not explain all these changes. Perhaps, as some argue, several important human pathogens mutated to milder forms. It seems that by 1700 the germ pool in the West was becoming domesticated, with microbes and humans evolving toward better tolerance; several deadly epidemics became milder endemics. In one way, it helped that more people lived in cities big enough to sustain all the crowd diseases. They were exposed early to many germs in tenements, factories, and schools, and with better general health and nutrition, many developed immunity without becoming seriously ill.

Vaccination was another check on infectious disease. For many centuries, people in parts of Asia, India, and the Arab world had practiced inoculation, scratching matter from smallpox sores into the skin of uninfected people. This usually caused a mild infection that created immunity to later exposures. Although inoculation sometimes caused sickness or death, the benefits exceeded the risks where smallpox was savage and widespread. Inoculation was introduced to England in the early eighteenth century and spread through the Western world. It was the military rather than civil governments which first made it mandatory. Washington had his army inoculated in 1776, and Napoleon did the same with his troops a few decades later.

At the end of the eighteenth century, inoculation was replaced by vaccination. The English physician Edward Jenner, having observed that milkmaids rarely bore smallpox scars, inferred that cowpox, a mild cousin of smallpox, was responsible. Vaccination, scratching infective cowpox material into the skin, was safer and more reliable than smallpox inoculation. Jenner's book on vaccination appeared in 1798. A decade later, the practice had spread to much of the world.

Vaccination was medicine's first major contribution to longevity and population growth. However, it could not become a model for preventing other infections until the germ theory explained how it worked. That happened late in the nineteenth century. By then, Europe's standard of living, life span, and population had advanced fitfully for a century or more. From then on, medicine, nutrition, and sanitation acted in synergy. Early in the twentieth century, infections were reduced further by better sewage treatment, chlorination, pasteurization, refrigeration, pure food practices, and pest control. These were instituted partly because the victory over cholera had encouraged government and society to accept responsibility for the health of nations and their citizens.

Around 1900, for the first time, cities could sustain their numbers without constant migration from the country. Most Western nations were becoming more urban than rural. Birth rates fell swiftly; the main reason was that cheap, effective contraceptives were available and in use. Yet populations kept rising. Families had only two or three children instead of five or ten, but most of them survived to adulthood. Of everything that contributed to better health and its consequences, medicine seemed the most dramatic and awesome. There is more glamor in discovering a germ than in filtering sewage or dating milk cartons. Perhaps drama and timing explain why medicine received more than its share of credit for the West's population growth and improved health and longevity.

By World War I, it was thought that diseases not yet conquered were lingering on borrowed time. Yet two dreadful pandemics, typhus and influenza, would kill tens of millions of people in half a decade, and medicine could cure or prevent neither. While typhus would be treatable with antibiotics a generation later, there is still no cure for influenza, and it remains one of the worst threats to human health and life. In 1918, when so-called Spanish flu struck, no one expected it to become a global peril.

The word "influenza" entered English from Italian in 1743. It meant influence—of a miasma, of the stars, or of both. Individual pandemics have often been named for their supposed origins, sometimes accurately (Asian flu) and sometimes not (Spanish flu). The infection's ultimate origin is farm animals. People have noticed for centuries that human outbreaks of flu often coincide with epidemics in pigs, ducks, and horses. Human flu probably dates no further back than their domestication, in the period 2000–5000 B.C. Influenza has never lost its link to its source. The virus owes much of its biological success to surface mutations that occur as it shuttles between humans and various domesticated animals.

Flu may have existed in ancient and medieval times, but the earliest undisputed accounts come from Europe's late Middle Ages, in 1387. Until 1492, potential pandemics probably died out on the continents where they arose. The first true pandemic probably occurred in the sixteenth century; there were five to ten more in the eighteenth and nineteenth centuries. Most of these began in Russia and Central Asia and traveled by land and by ship, even to remote Pacific Islands. The pandemic of 1833 was especially virulent; it laid low half the people of some European cities and killed countless tens of thousands. Perhaps because many of the dead had been old or already sick, there was no mass terror.

That is how influenza usually acts; it is one of the world's worst yet least feared killers. It causes a rather brief attack of fever, aches, and prostration, then a longer convalescence marked by fatigue and depression. The mortality rate is usually only about 0.01 percent, mostly from ensuing pneumonia. Yet so many people catch flu that despite a low death rate, the number of lives lost is often enormous. Since the immediate cause of death is usually a complication, the impact of the disease is measured by excess mortality, the number of deaths beyond normal levels in times of epidemic. In the United States flu results in tens of thousands of excess deaths even in its milder years. It is the only infectious disease that still ranks among the top ten causes of death.

The flu virus often undergoes minor genetic shifts in its surface proteins. Less often, there are major shifts; some researchers claim these happen in cycles of about sixty years, but others think they are random. All shifts make the virus an immunological stranger to humans, so in effect it is attacking virgin populations. That apparently happened in 1833 and 1889, when new strains of flu came out of Russia or Asia. The latter pandemic, the first to move with the speed of trains and steamships, killed 250,000 people in Europe alone. Despite its virulence, it had been almost forgotten in 1918, when the so-called Spanish flu appeared (again probably from the east). This pandemic was one of the worst disasters in history, and it holds puzzles for virologists and historians today. Their questions are more than academic. If another such virus should emerge—and many researchers expect it will—we may be little better equipped to fight it than people were in 1918.

From its first wave in the spring of that year, the Spanish flu was like none other. Rather than kill the old and sick, it slaughtered healthy young men and pregnant women. At crowded military bases, servicemen collapsed in feverish delirium, and many died of a raging pneumonia. Some doctors thought it was a new plague, and there were rumors of germ warfare by Germany. But it was flu after all, more malignant than at any time before or since, and often inviting secondary infections. With the second wave, in August, troop and supply ships spread the virus worldwide. It felled soldiers in every army, and civilians in every global region, from the tropics to the tundra.

Influenza killed viciously in South Africa, Siberia, and Samoa. In many American cities, half the people fell ill. Police and fire departments barely functioned; theaters, schools, libraries, churches, and pool halls were closed. Philadelphia was among the cities struck hardest. Flu and pneumonia killed 2,600 there in the second week of October, 4,500 the week after. New York City lost 9,000 in a single week. Doctors and nurses, many of whom worked bravely through the pandemic, perished in large numbers. The draft was suspended, troop ships were idled. There were scenes resembling those of medieval plagues— swamped hospitals, overflowing morgues, mass graves, and corpses in homes beside the sick and dying.

Every effort to prevent infection failed. People wore gauze face masks, in many cities required by law. They sterilized fountains hourly with blowtorches, wiped telephones with alcohol, and avoided shaking hands. Turning to hunches and folk medicine, they inhaled wood smoke, put sulfur in their clothes, had teeth pulled and tonsils removed. Doctors could suggest nothing better. Many mistakenly thought flu was caused by a bacterium that still is called *Haemophilus influenzae.* In Philadelphia and Boston, an alleged flu vaccine was distributed. Doctors and city officials knew it was useless, but they felt they should do something to raise morale and avert panic. The vaccine seemed to help, because flu was already fading as quickly as it had come.

Influenza deaths reported in the United States numbered 550,000, ten times the nation's death toll in World War I. Many cases went unreported; the real total may be as many as 650,000. One can only guess at how many died in such badly ravaged countries as India. The global mortality, usually given as 20 million, may have been 30 or even 40 million. World War I killed 15 million people in four years; flu killed perhaps twice that number in six months. Even bubonic plague did not kill so many people so fast.

It is astonishing that we did not all grow up with tales of the great flu disaster. It is equally amazing that there was little mass panic, and that the experience left only a light mark on history. An account of the pandemic by Alfred Crosby is appropriately titled *America's Forgotten Pandemic.* Only one important American literary work, Katherine Anne Porter's story "Pale Horse, Pale Rider," describes the social and personal experience of the pandemic in vivid detail. Today many Americans know more about medieval plague than about the biggest mass death of their grandparents' lives.

Crosby offers some speculations about why the pandemic seems sketched rather than etched in the world's memory. Perhaps, he says, the brutalities and passions of the war, then the worst in history, overshadowed the pandemic. Even if this is true, it cannot explain near amnesia over such a catastrophe. I do not know of one public memorial to the flu, its victims, or the doctors and nurses who died caring for the sick. There is, however, a monument to a much smaller medical crisis of the same era, and its presence is suggestive.

In Manhattan's Central Park, near East Sixty-seventh Street, there stands a large bronze statue of an Alaskan husky named Balto. He became an international celebrity for his part in the fight against an infection that has almost vanished in the West. Diphtheria, first described by the Roman Aretaeus in the second century A.D., was a common, dreaded infection that killed many adults and countless children by slow suffocation. Europeans brought it to the New World, and epidemics of "malignant throat distemper" swept colonial New England with mortality rates of 20 to 40 percent. Until a century ago, diphtheria was one of the worst epidemic terrors for American children. The germ, which acts through a powerful toxin, was the target of the first bacterial antitoxin ever used, in 1891. Today the DPT shot routinely protects against diphtheria, but outbreaks occurred in the West as late as World War II.

In the 1920s, getting antitoxin to diphtheria epidemics in out-of-the-way places remained difficult. Even small local outbreaks, when publicized, roused widespread public anguish over children's facing a painful and preventable death. That was the case in January 1925, when diphtheria broke out in Nome, Alaska. Nome was then a town of only 2,000, many of them natives who had never been inoculated against crowd diseases. Twenty-five children were sick with diphtheria, perhaps dying; dozens more had been exposed and were at risk. The nearest antitoxin was in distant Anchorage. Nome was storm-bound, and Alaska's only railroad came no closer to it than 650 miles. Newspapers around the world began to follow the attempted rescues.

Dogs, sleds, and antitoxin were shipped from Seward, north of Anchorage, to Nenana, the rail station nearest to Nome. In winter, the only way to travel the last 650 miles was the route the mails took, cross-country by relay teams of sled dogs. The trip usually took twenty days; the record was nine. The children of Nome did not have that much time. Dogsleds set out through a blizzard, in temperatures as low as 60 degrees below zero. Miraculously, they reached Nome with the antitoxin in five and a half days.

George Kasson, who drove the last leg of the trip, became an international hero. He said the real hero was his lead dog, Balto. The husky had created a trail to Nome through snow that often blinded Kasson. The trek is commemorated each year by a dogsled race that follows their route; it is named for Iditarod, a village that was similarly rescued in 1920. But nowhere was Balto a bigger hero than in New York City, where newspapers had headlined the story from its beginning. The parks commissioner raised funds for a statute of the dog, to stand in Central Park. Kasson and Balto traveled to New York for the dedication ceremony, which was nearly disrupted when Balto smelled another husky in the crowd.

The 1918 flu pandemic continues to recede from memory. Curiously, medicine was not blamed for failing to prevent 50 million deaths from flu and typhus in the world's last huge pre-AIDS pandemics. It, and Balto, remained heroes. Balto's memorial still stands; people stop to look at it every day, and to read the plaque describing his dash to Nome with antitoxin. The middle of his back is shiny where children have climbed onto him for seventy years. It seems that in the 1920s, the country saw its present and future not in the unsolved, lethal forces of typhus and flu but in the rescue of children from infectious disease. The implicit faith and optimism of their choice would last only decades.

a garden
of germs

The perils of prosperity. Polio and AIDS.
Germs do not forsake us.
Ancient triangles; humans stumble in.
Fevers and a striped invader.
Bolivia, Baltimore, and Shiprock.
The danger of deer mice.

If you are over fifty, you remember the summer terror of polio. If you are younger, you have known nothing quite like it. Each year, when the first cases were announced in newspapers and on radio, parents began the summer litany, warning their children away from swimming pools, movie theaters, and crowds. Poliomyelitis, like evil lightning, might strike anyone, anywhere, but it crippled and killed mostly the young. Hence its popular name, infantile paralysis. There was no treatment, not even a diagnostic test. From midsummer till fall, if a child came down with fever and sore throat, one could only wait to see if polio's stiff neck and aching muscles followed. If they did, the vigil began—for recovery, lifelong paralysis, or death.

From the 1930s on, dread of polio was stoked by the March of Dimes. To raise polio research funds, it blanketed the nation with heart-rending pictures of tots on crutches, their wasted limbs locked in braces. Newsreels reminded Americans of survivors imprisoned for life in the massive respirators called iron lungs. The only image of triumph over polio was President Franklin Roosevelt; stricken by the disease as an adult, he could stand only with ten pounds of steel supporting his legs.

At the turn of the century, polio was uncommon; known victims worldwide numbered only in the hundreds or few thousands a year. However, the number kept rising until the early fifties, when there were some 60,000 cases each year in the United States alone. Unlike most newly epidemic diseases, polio became not milder but more savage, leaving more of its victims dead or disabled. While one childhood disease after another was being cured or prevented, polio defied researchers. They ignored the evidence that it was a new kind of disease, rising not from dirt and dearth but from cleanliness and prosperity.

Some historians have claimed that polio goes back to ancient Egypt; it may, but the evidence is thin. The first undisputed case dates from the late eighteenth century, and for another hundred years victims were so few and scattered that polio was not thought to be contagious. The first recorded epidemic, a small one, occurred in rural Sweden in 1887. During the next twenty years, outbreaks there and in the northeastern United States had dozens or, at most, hundreds of victims. By 1907, the year of the first large epidemic, polio had appeared in prosperous nations from France to Australia. It was the big American outbreak of June 1916 that made polio a menace to every family with children. Some of the first cases turned up in clinics in immigrant neighborhoods in Brooklyn. In July and August, anxiety became panic; polio broke out in other boroughs of New York City, and spread through the East and Midwest. By year's end, there had been 27,000 cases and 6,000 deaths. New York City alone had almost 9,000 sick and 2,500 dead.

By ironic coincidence, that was also the year the American Museum of Natural History mounted pictures of 700 microbes in a display entitled "The Garden of Germs." The polio microbe was absent, but in that dawn of triumphs over contagion, many were confident it would arrive soon. In 1909, Simon Flexner of the Rockefeller Institute had learned that polio could be passed from monkey to monkey in the laboratory; the cause must be one of those invisible, newly hypothesized agents of disease, the viruses. In 1916, Flexner and his colleague Hideyo Noguchi told *The New York Times* that they had seen polio viruses through a dark-field microscope, and described them as "innumerable bright dancing points, devoid of definite size and form." Heaven only knows what they saw, perhaps dust or a contaminant; no virus is visible through an optical microscope. But the next year, a rendering of polio virus, perhaps based on the Flexner–Noguchi vision, joined "The Garden of Germs."

Any celebration was premature. There was still no polio test, no vaccine, no drug therapy. No one knew how the virus spread. Officials responded to polio's ravages as they often do to puzzling new challenges, by refighting their last war. Sanitation had helped conquer cholera and typhoid; many believed it would keep polio at bay. The question was what to sanitize. Many doctors thought the polio virus was airborne, coughed or sneezed into crowds like cold and flu germs. Playgrounds, movie theaters, and libraries were closed to children; some were shut down entirely. Perhaps the virus was waterborne; people were told to shun drinking fountains and swimming pools. Cities tried in vain to stop families from absconding with their children to presumed rural safety, so that they would not spread the disease. A number of local governments, including New York City's, required travel permits of people leaving from train stations, especially with children. Many citizens ignored the regulations and fled by any means they could find.

Perhaps, said some experts, physical touch and contaminated objects (fomites) spread polio. They scoured streets, parks, and schools, inspected water, milk, and food. Children were warned not to share toys. Streetcars and public phones were disinfected daily. People who had visited polio-stricken neighborhoods were advised to promptly wash their bodies, launder their clothes, gargle with antiseptics, spray germicide up their noses, and shampoo their hair.

Perhaps, other experts guessed, domestic animals spread polio; people abandoned or killed their cats and dogs. A more popular idea was that insects carried polio germs to people from garbage and from the horse manure still common in city streets. Patrick Manson had shown that mosquitoes carried malaria. Walter Reed then proved that they transmitted yellow fever; the discovery made him a national hero and enabled the building of the Panama Canal. This triumph of public health through insect control bolstered the speculation that houseflies might similarly spread typhoid and polio. Health departments urged people to seal garbage cans and screen their windows. Posters and pamphlets showed flies the size of rhinos menacing babies, and cities held fly-swatting contests for children.

Behind all the swatting, swabbing, and gargling lay a belief that dirt, and polio, were spread by the poor and foreign-born. Scientists and laymen alike feared that hordes of dirty, ignorant immigrants with primitive hygiene were infecting clean-living society. In 1906, in New York City, a poor Irish cook named Mary Mallon had become notorious as Typhoid Mary, a silent carrier who sowed disease and death in her wake. In 1916, the most recent immigrants, Italians and Slavs and Jews, were thought to similarly spread polio. Health workers aimed their efforts to clean, put up screens, and educate at ethnic ghettos in such places as Brooklyn, Manhattan's Lower East Side, and South Philadelphia.

Although some of the first polio cases in 1916 were reported by Brooklyn clinics, epidemiological evidence always pointed away from the urban poor. The earliest Swedish and American outbreaks were rural and suburban. In 1916, polio still preferred the clean, well fed, and uncrowded. Parts of Staten Island, the most suburban (in some places rural) of New York City's boroughs, had a population of only 2 people per acre; it had proportionally more polio than Manhattan, with up to 170 people per acre. More Staten Island victims, native-born and prosperous, suffered paralysis or death. The pattern held elsewhere; there was more polio on Philadelphia's prosperous Main Line than in the immigrant ghetto of South Philadelphia. Everywhere, polio was rare among poor blacks. Yet official attention stayed fixed on ethnic and racial slums.

Another myth, heightened by fund-raising publicity, was that polio remained a children's disease. For forty years after 1916, polio struck people at progressively older ages. By midcentury, some epidemics involved more adolescents and young adults than children. It also seemed to grow in virulence, crippling and killing more of its victims. As polio spread and attacked virgin populations around the world, those most at risk were between ages fifteen and twenty-five. Apparently polio had begun as a children's disease in the West only because most adults there were already immune.

Polio research made maddeningly little progress until 1948, when John Enders earned a Nobel Prize by devising ways to culture the virus in a test tube rather than in live monkeys. Now the virus could be produced copiously for study under the newly improved electron microscope, and its natural history would become clear.

The polio virus lives only in people; it probably adapted to the human small intestine countless millennia ago, transmitted perhaps from mice. Like many intestinal viruses, it thrives in warm weather and spreads by fecal-oral transmission; people pass it on with unwashed hands, directly or through fomites. After the virus enters the mouth, it multiplies in the throat and then in the intestine. Normally it leads to trivial illness or none, stimulating lifelong immunity to further exposure. But for reasons not fully clear, in a minority of people the virus travels through the bloodstream to the spinal cord and brain. The inflammation it causes there brings temporary or permanent paralysis; if the paralysis extends to vital organs, the victim dies. The later in life one meets the virus, the higher the chance that neurological damage will occur, and the greater the odds that such damage will be permanent.

This explains why researchers failed for so long to find polio's epidemic trail. The infection was everywhere; only exceptional cases were paralytic. The virus, once ubiquitous and silent, became a problem in advanced nations when poverty and dirt dwindled. Researchers found that in warm, crowded cities, such as Cairo and Bombay, most children were exposed to polio virus and immune to further attack by age three; paralysis and death were rare. This must have been true in the crowded, unhygienic cities of the West until the late nineteenth century. Then sanitation and hygiene improved, suburbs burgeoned, and crowding eased. Now many children did not mingle and exchange pathogens until they entered school, around age five. By then, severe symptoms were already more likely than in infancy or toddlerhood. All over the world, as prosperity and cleanliness increased, so did paralytic polio, at higher ages and with greater devastation.

Polio epidemics were still getting bigger and deadlier when the Salk vaccine was introduced, in the mid-1950s. Soon after, the Sabin oral vaccine appeared. These vaccines were so effective that today most doctors in developed nations have never seen polio. The last natural outbreak in the United States was in 1979. There have been no cases in the Americas for more than three years. Since the virus has no animal reservoir, worldwide eradication is theoretically possible. The World Health Organization (WHO) set 1990 as a goal for eradication of the virus, failed, and reset its target for the year 2000.

In the early 1990s, there are still more than 100,000 cases of polio each year, most of them in Africa, India, and China. In some places, vaccine is scarce; in others, it spoils for lack of refrigeration. In hot, moist climates, vaccine sometimes lacks full effect; perhaps flourishing intestinal viruses outcompete it in the body. Mysteriously, polio has sometimes broken out in vaccinated people in such places as Oman and Brazil. A better vaccine is probably needed, along with more funds and an aggressive vaccination campaign. But despite a failed first target date, eradication does seem possible.

The history of polio has some parallels with that of AIDS. Although the number of victims is smaller, and the disease less often fatal, polio evoked the helpless terror aroused earlier by bubonic plague and later by AIDS. (Unlike AIDS, it put the entire population at risk, regardless of behavior.) As with AIDS, no vaccine could be found, while cases soared; preventive measures aimed not at the microbe but at people's behavior, with limited results. There was much mistaken flailing in the search to understand transmission. For decades, polio drove virology research, just as AIDS drives it today. One must hope for positive parallels in the future. Long, intensive study eventually produced a polio vaccine, and so much was learned in the process that the polio virus was the first to be created outside a living cell, pieced together in a laboratory from its protein components. The present wave of AIDS research will probably have similar direct and indirect payoffs.

Eradicating polio is conceivable only because of the precedent set with smallpox. Smallpox virus, like polio virus, does not mutate often and has no animal reservoir, and one infection confers lifelong immunity. In 1967, when the WHO resolved to eradicate smallpox, vaccination had banished it from North America, Europe, China, Japan, and Australia; it lingered in thirty-one developing nations. No human pathogen had ever been wiped out, and some people thought the effort with smallpox futile and unsafe. Health workers labored for ten years in jungles and deserts, often among hostile populations; they had to improvise inspired solutions to unexpected technical and cultural barriers. In October 1977, in Somalia, they reported the world's last naturally transmitted (rather than lab-induced) case of smallpox.

Governments began to destroy their stores of smallpox virus and vaccine. However, real and theoretical risks of infection remained—the release of ancient smallpox virus from human remains preserved in permafrost or archaeological sites, accidents in labs, natural disasters, and biological warfare by governments or terrorists. In 1995, the world's last ampoules of smallpox virus remain in heavily guarded labs in Moscow and Atlanta, frozen in liquid nitrogen. Russia and the United States have agreed to destroy the remaining virus now that its genetic map has been completed; the deadline has been repeatedly postponed because virologists want to continue studying the microbe. If they do destroy it, this will be the first time any species has deliberately been made extinct.

In the decades following World War II, the campaigns against smallpox and polio suggested that people could rid the world of epidemics. A global war had been fought without a single pandemic. Antibiotics cured infections that once could be answered only with fatalism. Vaccines and sanitation slashed infant mortality. New pesticides controlled insect-borne diseases. Populations soared, but there was enough food to keep them alive and healthy. New herbicides and fertilizers increased farm productivity in what amounted to another Agricultural Revolution.

A complacent view arose, in which disease and health were items in a double-entry ledger. Microbes lay on one side, drugs and vaccines on the other. Soon, it seemed, each disease would be matched by a cure or preventive and thus canceled. The list would dwindle and finally disappear. Medical science's agenda would consist of cancer, cardiovascular problems, genetic disorders, senility, longevity itself. This was not a tabloid fantasy but the faith of many doctors and researchers. Unfortunately, their educations had been short on history, evolution, and ecology. Otherwise they would have doubted that human life could ever be divorced from that of microbes.

Indeed, pathogens were not forsaking humanity, but starting another assault. For each infection cured or controlled, many were emerging. They began to appear quietly in the peak polio years, from the thirties to the early fifties. Then more and more erupted, especially in the tropical forests and savannas of South America, Africa, and Asia. These places, with their lush diversity of species, were being transformed. The population growth permitted by drugs and pesticides was forcing people to plow new land, seek timber and metals, and build homes, highways, and dams. As during the Neolithic dawn, they were breaking ecological webs and stumbling into the territories of unfamiliar pathogens. They did so on the biggest scale in history; new epidemics and their carriers could circle the world at record speeds.

One might ask whether we have to review these emerging epidemics of the past half-century, some of them frightening, some foreign to wealthy nations and temperate climates. The prospect can induce a mild fit of medical student syndrome, but this new garden of germs calls for at least a brief stroll. Some of the infections seem safely exotic and distant, but many are on our doorstep, and some are already among us. Understanding their origin helps demystify such epidemics as Lyme disease, hantavirus infections, and AIDS, and warn us of others that are bound to come in the future.

Many of these diseases are caused by arboviruses, those transmitted by such arthropods as insects and ticks. Most of these viruses have ancient, symptomless relationships with the arthropods, which have passed them on to birds, rodents, and ungulates. Quite a few of these secondary hosts also came to tolerate the viruses, with few symptoms or none. They constitute reservoirs that reinfect the arthropods, creating a stable triangle of virus, vector, and reservoir. The triad many have some very specialized requirements. For instance, a mosquito can pass on only a virus it carries in its salivary glands, from which it is transmitted during feeding. The mosquito must linger in the right place at the right hour to reach its blood meal in a bird, monkey, or antelope. If it were not for such specialized adaptations, many viruses would leap with frightening ease from one species and region to another, devastating virgin hosts. When people break an ecologic web and intrude on ancient triangles of virus, vector, and reservoir, the results can be sickness or death.

The majority of arboviruses do not infect humans. In the early 1930s, only a handful were known, and just one, yellow fever virus, was thought to affect people. Now the count of arboviruses is more than 500 and rising; almost a hundred cause human illness. Some have been around for a long while, sporadically making people sick, and now are epidemic for the first time. Others have evolved or adapted to people quite recently.

Many arboviruses cause encephalitis, hemorrhagic fever, or both. Viral encephalitis ranges from mild to lethal. In the United States, many victims are forestry workers, campers, and children who play in woodlands, but the infection can strike anyone. While the number of victims is relatively small, the disease can be so deadly that even small epidemics cause panics.

Viral encephalitis usually begins with fever, headache, and a stiff neck; then inflammation of the brain can bring about lethargy, seizures, and death. Survivors may be left paralyzed, retarded, deaf, or blind. In North America in the 1930s, several forms of the disease were discovered when they attacked horses and humans. It soon became clear that encephalitis was actually a family of diseases, caused by related viruses, each spread by a different mosquito from reservoirs in birds or rodents.

St. Louis encephalitis (SLE) appeared in 1932. Eastern equine encephalitis (EEE) and western equine encephalitis (WEE) were first seen in humans in 1938. LaCrosse encephalitis, the most common type in this country, was not discovered until the 1960s. Of the four, EEE is rarest and deadliest; it can kill up to 80 percent of its victims and leave half the survivors brain-damaged. WEE is not as lethal, but it is much more common; epidemics have struck thousands of people from Canada to Argentina, with a mortality rate of about 20 percent. SLE and LaCrosse encephalitis take similar tolls. A single case of any viral encephalitis provokes an emergency response, including widespread spraying with insecticide.

Genetic studies suggest that many encephalitis viruses evolved in recent centuries, during a time of alternating deforestation and regrowth that altered the habitat, and habits, of many mammals, birds, mosquitoes, and humans. Irrigation may also have played a part; still or slow-moving water offers the mosquito that carries SLE virus a good breeding ground.

Where similar environmental changes occurred around the world, other types of encephalitis have emerged. Venezuelan equine encephalitis, first reported in 1938, is relatively mild. Rocio virus, which appeared in Brazil in 1975, is more dangerous. Deadly Japanese encephalitis, though now controlled by vaccines in Japan, is spreading in mainland Asia. There are related viral diseases, such as West Nile fever in Eurasia and Africa, Murray Valley encephalitis in Australia and New Guinea, Sindbis in Eurasia and Australia, and Russian spring-summer encephalitis in Central Europe. Most of these are carried by mosquitoes from reservoirs in birds, and some probably are capable of spreading beyond their original homes.

Encephalitis can be one symptom of the other major group of new arbovirus diseases, the hemorrhagic fevers. These are among the deadliest human infections, and their effects constitute a frightful catalogue. First come fever, headache, and pains in the muscles, joints, and abdomen. There may be a rash caused by small hemorrhages under the skin, which give the fevers their name. The victim's blood clots, yet it thins and leaks through blood vessel walls; it flows from internal organs, the nose, mouth, rectum, and even the eyes. The kidneys falter and may fail. Inflammation and bleeding in the brain can bring stupor, psychosis, seizures, or stroke. Death finally comes from bleeding and shock. Some hemorrhagic fevers are arthropod-borne, but others reach people from contact with infected animals or by inhaling microbes from their wastes. Some of the viruses can persist in human semen for weeks and may be sexually transmitted.

The best-known hemorrhagic fevers are yellow fever and dengue. Both have been spread by human activity from their original home in tropical West Africa; both have changed because people altered their lifestyles and environments, forcing the viruses to readapt. On each of these points, they are models for understanding the other hemorrhagic fevers.

The natural sources of yellow fever virus are monkeys in the high jungle canopy and a mosquito, *Aedes aegypti*, that breeds in tree holes. Sometime in the distant past, people picked up the virus from mosquitoes, and eventually a direct human-mosquito cycle of transmission developed in villages. Many people in endemic areas become immune through mild childhood infections, but yellow fever is deadly to newcomers; it was one of the reasons West Africa was called the white man's grave.

European ships carried the virus worldwide, through *Aedes* larvae in water barrels. For some reason, it never took hold in Asia, but by 1650, it had entered Latin America; there it thrived and was often fatal to both humans and monkeys. Ships carried it from Caribbean ports as far north as Boston. In 1878, yellow fever killed more than 20,000 people in a hundred American cities. After Walter Reed discovered its path of transmission, mosquito control reduced its prevalence. The last U.S. epidemic occurred in New Orleans in 1905. Eventually a vaccine was developed, and yellow fever became sporadic in much of the world.

Early hopes that the disease could be eradicated were dashed by puzzling urban epidemics, which were finally traced to an unexpected source. Loggers cutting tropical timber brought down trees, and with them came swarms of mosquitoes. Infected by insect bites, the loggers carried the virus back to towns, where they passed it to *Aedes* species that were able to breed in discarded tires, cans, and bottles. Modern villages had replaced treetops as *Aedes* breeding grounds—new homes for yellow fever, repeatedly renewed from jungle sources.

Insecticide sprays worked well against yellow fever for a while after World War II, but control lagged in the sixties. Today the disease claims victims in South America, and huge epidemics kill tens of thousands at a time in Africa, especially when war or civil unrest disrupts insect control. In 1992, yellow fever appeared for the first time as far east in Africa as Kenya. Whenever control measures flag, it will resurge in much of the tropical world.

While jungle and urban yellow fever are clinically identical, a similar difference in the sources of dengue has created a new disease more dangerous than the original. Dengue virus may have evolved from yellow fever virus; normally its effects are similar but less severe. In its classic form, dengue is rarely fatal, but the symptoms are excruciating— fever, chills, headache, diarrhea, nosebleeds, and muscle and joint pains that account for the name breakbone fever. Where dengue is endemic, it strikes mostly children, but in virgin populations it hits adults as well.

The natural history of dengue is much like that of yellow fever. It began in *Aedes* mosquitoes and African mammals, entered a direct human-mosquito cycle in villages and towns, and was spread globally by ships. It took hold in tropics and subtropics everywhere, including Asia. Dengue was rampant in the southern United States until the 1940s. The last major New World epidemic of classic dengue occurred in Trinidad in 1954, but it has returned sporadically in Latin America—for instance, in Costa Rica in 1993 and Puerto Rico in 1994, when it also struck U.S. soldiers in Haiti. Dengue is still epidemic in Africa, China, Southeast Asia, and Australia.

In the 1950s, a vicious new form of the disease burst out in Southeast Asia and the Philippines. Dengue hemorrhagic fever (DHF) could cause internal bleeding and lead to dengue shock syndrome (DSS), with coma and death. Fatalities sometimes ran as high as 15 percent. In 1981, DHF/DSS spread beyond Asia, to Cuba, where there were almost 350,000 cases in six months. Late in the eighties, it broke out in Puerto Rico and South America; now there are epidemics in the Caribbean almost every year. In the United States, Centers for Disease Control monitors every case of suspected dengue, fearing its return in a malignant new form.

Dengue took this lethal turn because there are four varieties of the virus, and infection by one does not protect against infection by others. Such multiple infection seems to send the immune system into destructive overdrive. Multiple infection probably began in the Philippines, where the physical havoc and population movements of World War II spread once isolated types of the virus and allowed mosquitoes to thrive. It has increased as both human and *Aedes* populations have exploded, with the mosquito thriving in poor, sprawling cities strewn with new breeding sites. Travel and trade have carried all four strains of the virus beyond their old boundaries. As a result, more of the world's urban poor have been exposed to at least two strains of the virus. DHF/DSS has afflicted a million children and killed tens of thousands. Conditions are ripe for its appearance in North America because of a newly imported carrier, the Asian tiger mosquito.

Aedes albopictus is called the Asian tiger because of the stripe down its body, but it deserves the name for other reasons as well. A prolific breeder in suburbs and cities, it is an aggressive daytime feeder that will attack almost anything warm-blooded. It thrives in discarded tires and tin cans, flower pots, birdbaths, even potato chip bags. The mosquito has long carried dengue in Asia and Hawaii. It entered the United States in 1985 in the puddles in used tires, which are shipped from Japan to Texas for recapping in huge quantities. When it arrived, biologists feared it might quickly outcompete less aggressive native mosquitoes and spread many diseases. They were soon proved right.

Despite control measures, the Asian tiger mosquito has reached more than twenty states, traveling as far north as Illinois and Delaware. Laboratory studies show it can carry yellow fever and six kinds of encephalitis. In 1992, a tiger mosquito was found in Florida that carried the deadly EEE virus. The species threatens to create epidemics of DHF/DSS, encephalitis, and even viruses not previously found in humans. It reminds us that new epidemics depend not only on microbes but on their carriers and on the environmental impact of human activity.

DHF/DSS was just one of many new hemorrhagic fevers to appear after World War II. Four occurred in South America, all linked to ecological disruption. The first came in 1953, in the Argentine pampas, where maize harvesters came down with fever, aching muscles, and bloody urine. In some places, one-fifth of those afflicted died. It took five years to find the cause, the Junin virus, and to understand why the epidemic had broken out.

During World War II, Axis and Allied powers both turned to Argentina for grain and beef. To meet the demand, virgin lands were plowed. After the war, new herbicides were used to control wild grasses that competed with the crops. The new landscape of cornfields and farmhouses offered endless bounty to *Calomys musculinus*, a fieldmouse that harbors the Junin virus. Usually the mice show no symptoms, but when their population surges too high, the virus brings it crashing down. Healthy or sick, the mice excrete Junin virus in feces and urine, and people inhale it with dust. Argentine hemorrhagic fever still occurs, and it is gradually broadening its range. Outbreaks peak every three to five years, along with the mouse population.

Junin virus worried few people outside Argentina, but it terrified researchers. It is a Biosafety Level 4 germ, the class of lethal microbe for which there is no vaccine and no cure. People who study it, and they are few, wear space suits and respirators, and work in maximum-security labs with negative air flow to keep germs from escaping. There are fewer than a dozen such labs in the world, and only two in the United States. They were needed when more class 4 viruses appeared in South America's changing ecosystems.

In the dry plains of eastern Bolivia, a sparse human population once raised stringy beef cattle. In 1953, a social revolution opened the area for farming; as in Argentina, new herbicides and pesticides made marginal land productive. The landscape was transformed, and human numbers swelled; so did the numbers of *Calomys callosus*, a relative of the mouse that carries Argentine hemorrhagic fever. When the mouse population became too dense, a usually symptomless virus the animals carried and transmitted sexually would cause a population crash. When mouse excreta accumulated around new farms and villages, the human population also fell. The germ that did the damage, a cousin of Junin virus, was named Machupo virus.

Bolivian hemorrhagic fever first appeared in 1960. It was like the Argentine fever, but even deadlier. When the epidemic peaked, in the mid-sixties, there were a thousand cases a year and hundreds of deaths. In one village, the ecological source of the disease became obvious. Spraying with DDT to prevent malaria had wiped out the village's cats; mice multiplied, and human illness followed. Destroying the mice ended the epidemic in precisely two weeks, the virus's incubation time. Surveillance and the trapping of mice have limited Bolivian hemorrhagic fever; there have been only about a hundred cases in the past twenty years. But if anything ever interferes with rodent control, the fever could explode again.

The next new hemorrhagic fever was a severe but seldom deadly infection called Oropouche. This time the germ was discovered before the disease. In 1951, the Rockefeller Foundation Virus Program set out to survey arboviruses around the world. During its twenty-year labor it discovered fifty new viruses in Amazonia alone. Everywhere it took the unusual and farsighted approach of seeking not only human disease viruses but those in healthy animals and insects as well.

In 1960, a Rockefeller team spotted a dead sloth on the new Belém-Brasilia highway and checked it routinely for viruses. They found a new one with no known link to human illness. It turned up again a year later, when a flulike epidemic sickened more than 11,000 people in Belém. The vector was discovered in 1980, when Oropouche struck 200,000 people in Amazonia. The culprit was a midge that normally fed on sloths and monkeys. Again the trigger for a new human epidemic was people's alteration of their habits and surroundings.

In the 1950s, highways had opened much of Amazonia to small farmers; needing a cash crop, they planted cacao. Soon mountains of cacao bean hulls rose around the villages. The hulls collected little pools of water that were perfect for breeding midges. The midges infected villagers with Oropouche; village mosquitoes may then have caught the virus from humans and spread it further. Huge epidemics of Oropouche have continued to seep through Amazonia. The first outbreak beyond

Brazil took place in 1989, in Panama. The story of epidemic Oropouche is still in the making.

A fourth South American hemorrhagic fever arose in 1989, among Venezuelan farm workers. At first it was taken for DHF/DSS, but most of the victims were adults, and the death rate, one in four, was too high. Early in 1991, researchers realized they were dealing with a rodent-borne agent related to the Junin and Machupo types; they named this one Guanarito virus. It remains to be seen what course Venezuelan hemorrhagic fever will take, and whether more such diseases are waiting to emerge in Latin America.

As these new hemorrhagic fevers appeared, environmental changes elsewhere in the world exposed people to other new microbes. In 1955, in the Kyasanur Forest region of Mysore, in southwest India, people noticed that bonnet monkeys were dying. As often happens, an animal epidemic heralded a human epidemic. The next year, the "monkey disease" struck people; it was a hemorrhagic fever that killed 1 victim in 10. Perhaps the virus was a new mutant; more probably, it had long circulated in ticks and rodents. It probably reached people because of deforestation to create new grazing land. The disease is still endemic in Mysore, but it has spread no farther. As DHF/DSS and Oropouche have shown, though, there is no guarantee that such diseases will remain local.

The worst new diseases of the postwar era appeared in Africa. There, too, the usual cause was a growing population that exploited the environment and invited zoonoses. Rift Valley fever (RVF) had been known in eastern and southern Africa since the 1930s, but chiefly as a disease of cattle and sheep. On rare occasions it resulted in encephalitis in people who worked with animals, and led to blindness or death. Suddenly in 1977, RVF struck hundreds of thousands of people in Egypt's lower Nile valley, where it had never been seen, and thousands died.

To this day, no one knows how the virus reached Egypt and became epidemic in humans. Perhaps it was carried there by herds driven to new grazing grounds, as human and animal traffic increased across Africa. Perhaps completion of the Aswan High Dam in 1970 offered a huge breeding site to the *Aedes* mosquito that carries RVF virus. A second epidemic, along the Senegal River, almost surely resulted from a new dam. In 1987, a year after the Diama Dam was finished there, RVF broke out in Mauretania and killed one-fifth of its thousand victims.

The virus must have lurked in an animal reservoir and increased as mosquitoes bred in the new expanse of water.

There have been no epidemics since, but RVF is so deadly that controlling it remains a major concern. That is no easy task, for the virus can survive for years in arid country, in drought-resistant *Aedes* eggs. When rain finally falls, it accumulates in shallow depressions, or dambos, and mosquitoes hatch forth to spread the virus. Satellite sensing imagery is being used to detect moisture patterns that favor *Aedes*, in order to pinpoint locations for preventive spraying. This technique, though expensive and unproven, is also being explored in hopes of predicting and preventing outbreaks of yellow fever and malaria. So far, RVF virus has not appeared outside Africa, but mosquitoes in Europe and North America can carry it.

Three other lethal viruses emerged in Africa, and all sent waves of alarm around the world. The first, Marburg virus, was spread by the international trade in lab animals. Virus research and vaccine production had created tremendous demand for monkeys. In 1967, green monkeys shipped from Uganda to polio vaccine plants in Europe began to sicken and die. Soon animal handlers and lab workers in Marburg, Germany, came down with a raging hemorrhagic fever. Some passed the infection to doctors and to relatives who cared for them. No attempt at treatment helped, and 7 of the 30 patients died miserably.

The movement of any infection between continents makes health officials anxious, and this one was murderous. The United States and several other countries enacted long, strict quarantines on imported monkeys. Although the epidemic ended, it left ominous questions unanswered. What was the source of the virus? It was lethal to monkeys, so it was probably as new to them as to people. Years of searching for the microbe's African reservoir have been fruitless. How is the virus transmitted? One Marburg victim caught the virus from another's semen, and a few were infected by other patients' body products. Yet neither of these seems to be the usual route of transmission, which remains unknown.

There have been no more epidemics of Marburg fever, but several cases occurred in Africa in the 1980s. Freeze-dried specimens of the Marburg virus are stored in a few high-security labs around the world, awaiting study if there is another epidemic. The virus has already proven it can cross the boundaries of species and of continents. The question is whether, or when, it will do so again.

In 1969, another hemorrhagic fever emerged, at a mission hospital in Lassa, Nigeria. The victims suffered fever, encephalitis, seizures, hemorrhage, and shock. Two of three infected nurses died. So did one of two American lab workers, infected by specimens from victims' bodies. The disease recurred in epidemics for years, and there were scare headlines around the world when infected tourists carried the virus by plane to London, Toronto, and Chicago. Mercifully, none of these cases led to further epidemics.

The danger of Lassa fever is illustrated by a case that received wide publicity. In 1989, a man flew from Chicago to Nigeria for his mother's funeral; she had died there of an undiagnosed fever, as had her husband ten days earlier. Soon after returning to Chicago, the man went to a clinic complaining of fever and a sore throat. A flu epidemic was in progress, so he was told to go home and take acetaminophen. He returned to the clinic three days later, much sicker, and received penicillin. Five days later he was back again, with a swollen face and bloody diarrhea. An alert specialist suspected hemorrhagic fever and contacted the CDC. The Centers' diagnosis of Lassa fever came too late; the man was buried two weeks after his return from Nigeria. Since then, he had been in contact with more than a hundred relatives and health workers. Each one was tracked down and given antiviral medication. Miraculously, not one was infected. Still, the case is a reminder of why hemorrhagic fevers are on the world's medical alert list.

The source of Lassa fever has been discovered in a small mouse that scavenges in and around dwellings in West Africa. The virus, excreted in mouse urine, infects people through breaks in their skin or by their contact with human victims' blood, tissues, or excretions. In some parts of Africa, more than half the population has antibodies to Lassa virus, which means they were infected without showing major symptoms, perhaps in childhood. But when the virus does cause disease, it can kill one-fifth of the sick. The reason for every outbreak is not known, but sometimes the trigger is clearly environmental upheaval. In Sierra Leone, for instance, Lassa fever epidemics followed the chaotic development of surface diamond mines, with a sudden influx of people, primitive boom towns, and feasts for scavenging rodents.

The range and variety of Lassa-type viruses is greater than virologists first suspected. Two less virulent strains, Mozambique virus and arenavirus 3080, have been found, and others probably await discovery. There is now drug therapy for Lassa fever, and work proceeds on developing a vaccine. While rodent control has helped limit the disease in Africa, there are thousands of cases each year, and many deaths. The risk remains that one soldier, tourist, or businessman leaving Africa could spread the virus not only to other people but to some new animal reservoir in another part of the world.

In 1976, a new virus appeared that was even more frightening than Marburg and Lassa. Ebola fever, the deadliest human infection besides rabies and AIDS, broke out suddenly in Nzara, a town in southern Sudan. At first, victims had fever and joint pains; then came hemorrhagic rash, kidney failure, seizures, bleeding, shock, and death. The virus spread to a hospital in a nearby town, where it killed many of the patients and staff, and then to relatives who had cared for the sick. Of almost 300 victims, more than half died.

Two months later, the fever broke out 500 miles away, in the Ebola River region of Zaire. It killed 13 of 17 staff workers at a mission hospital and spread to patients, who carried it to more than fifty villages. One nurse, fighting for her life, was taken for treatment to the country's capital, Kinshasa. This poor, crowded city of 2 million had direct air links with Europe. Fearing a pandemic, officials had the Zairian army seal off the Ebola region, and an international scientific team was rushed to Kinshasa.

Among the team was virologist Karl Johnson, who had studied the first Machupo virus outbreak. Johnson later said that on learning the death toll among the sick in Zaire approached 90 percent, he and his colleagues feared they might be facing an Andromeda strain that could kill millions. Instead, the epidemics subsided in Zaire and Sudan as unexpectedly as they had begun. In 1979, another outbreak hit Nzara, and scattered cases occurred in the 1980s. By then researchers had learned something about the disease, but less than they would have liked.

Ebola virus, like Lassa virus, is endemic in parts of Africa, usually causing mild symptoms. Although none of the epidemics originated in a hospital, all were first spread in clinics through reuse of hypodermic needles and patient contact. To this day, the reservoirs and transmission of Ebola and Marburg viruses are mysteries, and there is no drug or vaccine to combat them. The prospect of dealing with Ebola virus even in high-security conditions has been known to put researchers in a cold sweat. Although the Ebola, Lassa, and Marburg microbes faded from headlines when epidemics subsided, virologists feared that they might be warnings of future crises. It briefly looked as if that was happening in 1989, when an Ebola-like virus rampaged through laboratories just outside Washington, D.C.

In the late 1980s, 16,000 cynomolgus monkeys were being imported to the United States each year for research. The quarantine enacted after the Marburg epidemic had appeared successful in screening out dangerous pathogens. Then in November 1989, monkeys in a lab in Reston, Virginia, began to die. The lab staff suspected simian hemorrhagic virus (SHV), which is lethal to monkeys but harmless to humans. SHV was exonerated, though, and monkeys kept dying. Fear of a Marburg disaster in the making prompted a call for help to the U.S. Army Medical Research Institute of Infectious Disease (USAMRIID) at Fort Detrick, Maryland.

Fort Detrick scientists confirmed that a previously unknown hemorrhagic fever virus was to blame. Workers in space suits promptly euthanatized all the monkeys at the Reston lab and performed a radical disinfection known as "nuking"—saturation with germicides and formaldehyde gas so formidable that no organism can survive. Several workers at the Reston lab developed antibodies to the virus, but none fell ill; the virus seemed not to sicken humans.

This was less than fully reassuring. A small mutation might make such a virus virulent to people. Reston virus, as this cousin of Ebola was named, raised nightmare speculations. No Ebola virus had been found in monkeys or outside Africa; these monkeys were from the Philippines. The reservoir was almost surely another species; that meant an Ebola-like virus was circulating in Asia in unknown hosts. There was potential for a global outbreak of a deadly zoonosis. The quarantine and testing of imported monkeys became more stringent than ever. So far it has proven effective, but until the source of Reston virus is discovered, complacency will be out of order.

The terrible trio of Ebola, Lassa, and Marburg fevers has touched few people beyond Africa.* But that could also have been said a dozen years ago about AIDS. Virologists monitor these and other viral infections—especially forms of encephalitis, hemorrhagic fever, and immune deficiency diseases—that have emerged or spread dramatically in recent decades. Most are zoonoses traceable to disruptions of environments or to novel human activity. Some have had far more victims than Ebola, Lassa, and Marburg combined. Among them are:

Crimean-Congo hemorrhagic fever

Carried by ticks from many ungulates, it can kill up to half its human victims. Though perhaps long present in Asia, it first gained wide attention in the West after a Crimean outbreak in 1944–45. A virtually identical but less lethal strain of the virus has been found in Africa, the Middle East, and parts of Europe. There is no drug therapy.

Omsk hemorrhagic fever

Recognized after World War II in Russia and Siberia, this dangerous tick-borne zoonosis resembles Kyasanur Forest disease.

Chikungunya

This hemorrhagic fever, which first appeared in 1955, is not usually fatal, but it causes excruciating joint pain. The virus reached people in East Africa probably because of increased contact with monkeys or with mosquitoes that feed on them. It was spread by ships to India, the Philippines, and elsewhere in Asia.

O'nyong-nyong

The pathogen may have evolved recently from the chikungunya virus; the disease appeared in 1959, for similar reasons, affecting 2 million Ugandans with fever and painful arthritis.

Ross River fever

The virus causes an epidemic arthritis that can last for weeks or years. Carried to people by mosquitoes from a reservoir in kangaroos, it was identified several decades ago in Australia and Southeast Asia. Irriga-

---

* As this book goes to press, another outbreak of Ebola fever has started in Africa and is receiving attention worldwide.

tion and dam building there had created new hatcheries for the carrier. Explosive epidemics struck virgin populations in South Pacific islands in 1979–1980.

Monkeypox

The disease surfaced in 1958 in lab monkeys from Africa. It first struck humans in 1970, infecting mostly village boys in Zaire. The germ, a close relative of smallpox virus, causes similar sickness in people, killing 1 in 10. The reservoir is squirrels and other small mammals; monkeys, like people, are accidental victims. Boys probably caught the virus by hunting, skinning, and eating small game. After peaking in the mid-1980s, monkeypox declined, probably because so much of Central Africa's rain forest had been destroyed.

Even many people who knew of these diseases considered them exotic and little threat, limited to safely distant ecosystems. There is no such complacency about hantavirus infections, which for forty years have been increasing around the world. Their extent and their danger to humans are only now becoming understood—especially since a deadly epidemic struck the United States in 1993.

Western medicine first met this family of viruses in the early 1950s, during the Korean War. Korean hemorrhagic fever had been described decades earlier, but its cause and extent were unknown. Some scholars think it has a long history in China; if that is so, it remained a local problem. In the first half of this century, it struck Korean farmworkers during rice harvest. Under war conditions, in the early fifties, it reached United Nations soldiers; several thousand fell sick with fever, kidney damage, bleeding, and shock, and 400 died.

The virus, isolated in 1976 and named for Korea's Hantaan River, reaches humans from the wastes of a fieldmouse it chronically infects. So far, this sounds very much like the story of Machupo, Lassa, and other rodent-borne hemorrhagic fever viruses. But the discovery of Hantaan virus proved to be only the first step in a story of medical detection that is still in progress.

Hantaan virus is more widespread than was first thought; it infects people in much of Asia and Russia. It and its relatives strike at least 200,000 people a year and kill up to 20,000. About half the victims are in Korea and China, infected by a mouse-borne relative of Hantaan virus called Seoul virus. Another relative, Puumala virus, causes nephropathia epidemica, a kidney disease. First described in Finland in 1930, the disease occurs from the Russian Urals into Europe. The virus, carried by wild voles, was isolated in 1979. Yet another relative of Hantaan virus, the lethal Porogia virus, was found in 1986 in the Balkans; it brings about a cruel hemorrhagic fever. The Hantaan, Seoul, Puumala, and Porogia types are now labeled hantaviruses, and the ills they cause are collectively called hemorrhagic fever with renal syndrome, or HFRS.

HFRS and the hantaviruses at first seemed to be exclusively Old World problems. Then in the late 1970s, the reservoir of the Seoul virus was found in rats, which carry many microbes worldwide on ships. When Seoul HFRS appeared in the port city of Osaka, uneasy researchers began testing rats for antibodies to the virus in ports around the world. Results were positive on every continent. In 1982, a new variety of hantavirus, the Prospect Hill strain, was found in wild voles outside Washington, D.C. It caused no disease in humans, but its existence provoked more hantavirus testing across the country. A few years later, antibodies to still another Seoul-type virus were found in rats in Baltimore. Rats infected with this Baltimore rat virus (BRV) flourished not only near the port but in the inner city's trash and litter.

The next logical question was whether BRV had been the source of undetected or misdiagnosed disease. A 1986–1988 study at Johns Hopkins Hospital turned up fifteen patients with BRV antibodies; many had suffered various combinations of kidney disease, hypertension, and stroke. Later, three patients with active BRV infections were found; all lived in a neighborhood where infected rats thrived. All had the symptoms of HFRS, and two developed chronic kidney disease. BRV antibodies were also found in patients at inner-city dialysis clinics in Baltimore.

Scientists wondered whether a new viral disease had been caught in the act of emerging or an old infection was being belatedly identified. The answer is not certain, but some hantaviruses of the Seoul variety were probably spread worldwide from Asia recently. New hantaviruses have been discovered in India and Thailand, and in New York, Philadelphia, Cincinnati, New Orleans, Houston, and other U.S. cities. More study is needed to determine which of these have spread lately from Asia or Europe, and which are native. Suspicion is high that hantaviruses, old or new, cause many cases of kidney disease, hypertension, and stroke.

Virologists' concern about hantaviruses was further justified in 1993, when a "mystery disease" broke out among Navajos in the Four Corners region where New Mexico, Arizona, Utah, and Colorado meet. It began abruptly with such flulike symptoms as fever, aches, coughing, and exhaustion. Although many victims were young and fit, they collapsed suddenly and went into respiratory failure. Their lungs filled up until they drowned in their own blood. No drug could help them.

One of the first victims was a teenage Navajo girl in New Mexico. On the way to her funeral, her boyfriend fell ill with the same symptoms. He died a few days later. A thirteen-year-old girl collapsed while dancing at a graduation party and died the next day. By late May, two dozen people were being treated for this raging pneumonia in hospitals in Albuquerque and Shiprock; ten others had died. *The New York Times* reported on its front page that a deadly new disease had suddenly appeared.

A team of expert investigators met in the Southwest on Memorial Day weekend. They noted that despite a few dramatic exceptions, the disease did not pass easily to victims' family and friends. Rather, it struck almost randomly over a huge geographical area. That suggested an animal source, perhaps with secondary person-to-person transmission. A virus inhaled from animal droppings might account for the galloping pneumonia. The team first checked for bubonic plague, long endemic in wild rodents in the Southwest. That possibility was eliminated. They checked for influenza, bacterial pneumonias, Legionnaires' disease, anthrax, mycoplasms, Epstein-Barr virus, HIV, and a variety of chemical and biological toxins. All the results were negative. Hantaviruses were on the suspect list, but far from the top; some virologists doubted they should be there at all. They had never caused such symptoms, nor had they ever been seen in this region. Then in early June, a new hantavirus was found in wild mice in New Mexico.

The handful of researchers who had worried about hantaviruses as a global threat could well have said "I told you so." Research on hantaviruses, never extensive in the United States, had actually been cut back. The army had studied them since the Korean conflict, but after the Persian Gulf war, Congress criticized such "arcane" inquiry at the expense of germ warfare research, and USAMRIID began shutting down its hantavirus projects. Fortunately, a small team had continued its study of hantaviruses at Fort Detrick and developed a diagnostic test. By mid-June, after the sixteenth death, the test confirmed that a new hantavirus was responsible for the Four Corners deaths. Meanwhile, rodents were being trapped and tested, and deer mice were identified as the virus' carriers.

Researchers might have spent years trying to figure out why this common scavenger was suddenly spreading the virus to humans. The answer, however, was at hand, for a New Mexico biologist, Robert Parmenter, had just finished a ten-year study of local rodents. Heavy rain and snow the previous year had caused a rare abundance of piñon nuts and grasshoppers, staples of the deer mouse diet. When the animals' food supply expands, they have bigger, more frequent litters, and their population grows. From May 1992 to May 1993, deer mice had multiplied tenfold. As a result, people were exposed far more than usual to the mice and their wastes. Late in the summer of 1993, the mouse population started to fall, and the hantavirus epidemic in humans waned. Nature had created a zoonosis through ecological change, and for the moment, at least, nature was letting it recede.

By August, the epidemic had killed 26 of the 42 reported victims, a death rate of more than 60 percent. It was imperative that research continue, in case the epidemic continued or returned. The new disease, named hantavirus pulmonary syndrome (HPS), had begun among Navajos, but it had also appeared in Hispanics and Anglos. The early talk of a "Navajo disease" had been wrong. In fact, since deer mice live in much of North America, the disease might be widespread. The CDC asked doctors all over the country to report cases that resembled HPS. It learned that 16 people had been infected in Texas, 6 fatally. Cases were identified in Colorado, Nevada, California, and Louisiana, and more reports trickled in during autumn, from places as distant as Florida and North Dakota. By year's end there had been 54 cases and 32 deaths.

The virus recovered from the HPS patient in Louisiana belonged to a different strain, so at least two new types of hantavirus were abroad. Since each strain tends to adapt to a particular species of rodent, several carriers might be involved. The HPS virus was found to be carried by at least a half-dozen rodent species, including the common house mouse.

Genetic studies of HPS virus showed that it is closely related to the

Prospect Hill virus. There was fear that perhaps a mutation had allowed the virus to leap to new carriers. That had happened before; in the Balkans, the Puumala hantavirus, usually carried by voles, had recently been spread to humans by the house mouse. This fact raised concern that rodents might cause more cases of HPS in winter than in summer, because mice often try to escape the cold by invading human dwellings.

During the winter of 1993–1994, studies of old blood samples from unexplained pneumonias revealed that although HPS was rare, there had been scattered cases as early as 1980. Now that researchers knew what to look for, they were finding it. HPS had occurred here and there, and usually was tossed into such wastebasket diagnostic categories as adult respiratory distress syndrome (ARDS). So far, it seems that most or all HPS infections were picked up from aerosolized mouse wastes, but virologists speculate that HPS could conceivably become an airborne pneumonia passed from person to person. In 1994, the microbe was ominously named Muerto (Death) Canyon virus; later this was diplomatically changed to Sin Nombre (no name) virus. By late in the year, the CDC had counted almost a hundred cases in twenty states.

New hemorrhagic fevers, appearing from the United States to Zaire, and from Argentina to Korea, have defied the West's recent faith in a germproof future. When most of these epidemics arose, it was tempting to see them as the distant miseries of impoverished nations. Even when Marburg, Lassa, and Ebola diseases made headlines, they still seemed to most people in developed countries like nightmares unrelated to their own lives. Only a few researchers, outside the mainstream concerns of science, saw that these new diseases represented a coming wave of change in human behavior and in ecosystems both local and global. People in developed nations began fitfully to sense that they were living in a new age of infections when Lyme disease, Legionnaires', and AIDS arrived.

an old thread, new twists

Old Lyme, ticks, and second nature.
Lethal hotels, toxic hospitals.
The third retrovirus. And more catapults.
Needles, sex, and the monkey trade.
Another AIDS?

Lyme disease, Legionnaires' disease, and AIDS are the new infections with which Americans are most familiar. They all arrived in the past two decades, each deadlier and more widespread than the one before. Although they differ in symptoms, causes, and impact, a common thread runs through their origins. Each arose when people changed their environment and their behavior, offering microbes new niches and paths of transmission. This is an old, familiar pattern in the emergence of diseases, but now it has a surprising twist. In the past, most new epidemics resulted from changes that degraded the environment, such as deforestation, expanded farming, and dam building. Lyme disease, however, emerged from a convalescing environment, Legionnaires' from new protected environments, and AIDS partly from changes meant to enhance life. They show that a shift, any shift, in ecosystems or in human behavior can tangle the web of life.

Lyme disease was the first of the three to appear. In 1975, two mothers in Old Lyme, Connecticut, concluded that something strange and dangerous had struck their families. Each had a child suffering long, recurrent bouts of fever and aching joints. They were diagnosed with juvenile rheumatoid arthritis (JRA), in which the immune system attacks joints as if they were foreign bodies. Although rheumatoid arthritis is rare in children, and these two had some atypical symptoms, no alarm went off in their doctors' minds.

The mothers did become alarmed when they learned that a dozen other children in the area had the same problems and diagnosis. That seemed to defy probability. Fearing an infection or environmental toxin, they called the state health department and the Yale Rheumatology Clinic. JRA is not infectious, they were told, and does not occur in clusters. A dozen cases in Old Lyme would mean it was occurring there at thousands of times its normal frequency.

The Connecticut children had all fallen ill in summer, in a suburban setting. That suggested an arthropod-borne disease, but their symptoms matched no known infection. Dr. Allen Steere, then of Yale, led a team that surveyed the adjoining towns of Lyme, Old Lyme, and East Haddam, seeking similar cases. The syndrome, they found, had afflicted 39 children and 12 adults in the past three years. In 1977, Dr. Steere published a description of a disease he called Lyme arthritis. It had the hallmarks of an infection, though no microbe had yet been found. Many patients had developed a rash when the illness started, and several recalled having been bitten by a tiny tick where the rash began. Fortunately, one of them had saved the offender, and turned it over to researchers. It was *Ixodes scapularis* (formerly called *I. dammini*), the deer tick. Field workers began collecting ticks around Old Lyme to see what germs they were carrying and to study the tick's life cycle.

"Deer tick" is a misleading name. The deer is merely a brief feeding and mating station at the end of *I. scapularis*' two-year existence. The tick hatches in spring as a six-legged larva the size of a pinhead. It feeds mostly on rodents, especially the white-footed mouse, from which it acquires the Lyme microbe. Over the winter, the larva changes into an unlovely eight-legged nymph. Still carrying microbes from the previous summer, it passes them to almost any warm-blooded host it finds—raccoons, squirrels, skunks, birds, or humans—but its favorite is still the white-footed mouse. At summer's end, the nymph becomes a tick, smaller than a printed period. That is when it eats its last supper and finds its mate, on the body of a deer. For lack of a deer, it may attach itself to a cow, horse, dog, cat, or human, and pass on the Lyme infection. All of these hosts suffer fever and arthritis as a result.

The germ was discovered in 1981, almost by accident, after Rocky Mountain spotted fever (RMSF) broke out on Long Island. The appearance there of this infection was less odd than its name suggests. RMSF first reached humans late in the nineteenth century, in Montana's Bitterroot Valley, killing up to 80 percent of the infected. Most victims were prospectors, miners, and settlers. Having deforested large areas, they worked in the resulting scrub landscape and caught from wood ticks a rickettsial organism related to the one causing typhus. The germ and its vectors are now widespread. RMSF, which has a natural reservoir in small rodents, occurs from Canada to Brazil. Outside the Rockies, it is usually less virulent, but still dangerous. A similar disease, ehrlichiosis, common in dogs and transmitted by the brown dog tick, first appeared in humans in 1986, and in some areas is more common than RMSF.

The CDC carefully monitors such infections as RMSF and ehrlichiosis. So when RMSF surfaced on Long Island in the 1970s, biologists sought the source of the outbreak by collecting ticks there. The specimens were sent for study to a Public Health Service laboratory in Montana that was experienced with tick-borne diseases. There Dr. Willy Burgdorfer dissected the ticks and cultured microbes from their bodies. In the deer ticks he noticed an unfamiliar member of the *Borrelia* family, a pale, slender spirochete related to the syphilis germ. He injected it into rabbits, and they developed the symptoms of Lyme disease. Burgdorfer received the reward of those who discover new species: the bacterium was named *Borrelia burgdorferi*.

Although its cause is known, Lyme disease remains a diagnostic nightmare. Only half of its victims show a telltale bull's-eye rash at the site of the tick bite. Most develop fever, headache, sore throat, nausea, fatigue, swollen glands, stiff neck, and aching muscles; at this point they may be diagnosed as having anything from flu to rheumatoid arthritis. A dozen laboratory tests for B. *burgdorferi* exist, but all can give false results. Many patients recover within days or months, even without antibiotics, but some become chronically ill. The spirochete can change its surface proteins to evade both drugs and the immune system. It is unclear whether symptoms sometimes persist because the germ hides in cerebrospinal fluid or because it sets off autoimmune responses that outlive the infection.

Like syphilis, Lyme disease can subside during a long latency period and then reappear to attack the joints, heart, or nervous system. Long-term cases present a bewilderment of symptoms that last for years. The arthritis can become so severe that it requires knee surgery. Heart damage may necessitate implanting a pacemaker. There can be such neurological complications as impaired memory, meningitis, or facial paralysis.

Lyme disease increased twentyfold over a decade, from 500 new cases in 1982 to 10,000 in 1992. It became a staple subject of television, newspapers, and newsmagazines. In hard-hit areas, fear turned into hysteria. The reaction was understandable; Lyme disease was increasingly common, often misdiagnosed, sometimes crippling, and it could drag on for years. Even where it was not common, many parents fearfully kept children indoors during the summer. The prevention and treatment of Lyme disease became an industry. Stores and mail-order companies peddled tick repellents and home diagnostic kits, none foolproof and some probably of little use.

Despite people's precautions, Lyme disease continues to become more common and widespread. It has appeared in almost every state; the greatest number of cases occur in the Northeast and upper Midwest and on the West Coast. Some experts think it is overreported, others that it is underreported. Even if underreported, it is now the most common arthropod-borne infection in the United States.

Lyme disease is not, in strict terms, brand-new, but it is novel as an epidemic. When Burgdorfer discovered its cause, he recalled a lecture he had heard decades earlier about a similar, milder condition reported sporadically in Europe. In Germany in 1883, and then in Sweden in 1910, doctors had described a rare skin disorder that sometimes caused neurological complications, including stroke. They suspected it was carried by ticks. Burgdorfer correctly guessed that these reports of "erythema migrans" (traveling rash) were early, scattered cases of Lyme disease or something like it. Still, it was puzzling that the infection had suddenly become severe and epidemic in the United States, with arthritic symptoms rarely seen in Europe.

Clues to how and when the disease had spread might lie in tick collections in science museums. Even in ticks pickled in formaldehyde for decades, tests could detect remnants of spirochetes' surface proteins. Traces of *B. burgdorferi* showed up in ticks collected on Long Island in the 1940s, and in other pre-1975 specimens from Cape Cod and Wisconsin. In Montauk, at the eastern tip of Long Island, people had spoken for years of an arthritis they called Montauk knee. It turned out to be Lyme disease.

Clearly Lyme disease had been around in this country for decades before 1975, undiagnosed or misdiagnosed. It or a precursor had existed in Europe. But it became epidemic only when changed environments and new lifestyles offered the microbe a new opportunity. The reason was not that people were exploiting the landscape but that they were helping it recover.

We saw earlier that what European explorers found in the New World was not virgin forest. For millennia, Native Americans had hewn, burned, and planted, reducing first-growth woodland in the Northeast. Then for three centuries, white settlers sacrificed forest for farmland, timber, and fuel. By the early 1800s, the East was almost treeless. This, as much as hunting, brought deer and their predators to the brink of extinction. During the nineteenth century, agriculture shifted to the Midwest; farms in the East shrank in size and number. Land began reverting to scrub and regrowth forest. At the turn of this century, federal and local governments acted to preserve and restore vast tracts of land. Forests and wildlife rebounded.

Patchy new woodland is ideal for deer; they thrive best at forests' edges. By 1900, the deer population of the East was recovering; the populations of predators such as wolves, bears, and big cats were not. The recuperating landscape allowed a resurgence of other wildlife, including rodents and ticks. And it drew another species, humans eager to flee crowded industrial cities. Urbanites who had never seen true wilderness mistook second- and third-growth woodland and scrub for forest primeval. They didn't really want raw nature anyway. They sought a touch of nature along with urban amenities. The result was a compromise—suburbia.

At the end of World War II, the East had several times as much forest as a century earlier. The deer population was exploding. All over the country, an urban housing shortage and the postwar economic boom combined to create vast swaths of suburban and semirural housing. Deer adapted to them as resourcefully as squirrels and raccoons. Today deer crop the grass in suburban yards, loiter in city parks, and are common highway hazards. Many exasperated homeowners and motorists regard them as vermin and refer to them as "rats with hooves"— especially where Lyme disease is rampant. People are learning that a convalescent environment creates its own peculiar problems. Ecological change of any kind creates a risk of human disease.

Since the 1980s, similar regrowth has increased in Europe, with flourishing deer, rodents, ticks, and suburbs, and big outbreaks of Lyme disease. Europe now has twice as many cases as the United States; Germany alone has tens of thousands each year. Lyme disease has been reported also in Japan, China, Australia, and South America. This may be just the start of its story as a global epidemic.

We still have much to learn about Lyme disease. Biologists suspect that it is spread to new areas by ticks on migrating birds. It is not certain why the European form is milder or why it tends to cause neurological rather than arthritic complications. Perhaps each strain of the germ has adapted to a different vector, developing slightly different characteristics. In Europe, Lyme disease is caused by the sheep tick; in California, its life cycle involves the dusky-footed wood rat and two species of ticks other than *I. scapularis*. Prevention and treatment remain problematic. There is an effective vaccine for dogs, to reduce transmission to pet owners; vaccines for humans are still being tested. Antibiotics help many patients, but sometimes they cannot wipe out the infection. The best protection is still using repellents, examining people and pets for ticks, and seeking prompt attention for tick bites.

Recovering woodlands have given humans another tick-borne disease. The babesia parasite, a protozoon that invades red blood cells, causes a malaria-like illness in cattle. On Texas ranches, where it used to do great damage, it was called redwater fever. It was never known to infect people until 1957, when sporadic cases appeared in Yugoslavia and then on Nantucket Island, an old whaling center off the Massachusetts coast.

On Nantucket, as in most of the Northeast, forests and deer had almost vanished by the nineteenth century. The trees had been felled to create grazing land for sheep; the sheep reduced the island to a close-cropped pasture, eliminating the home of both deer and ticks. Then pine trees were planted as windbreaks, sheep farming and whaling declined, and cranberry farming and summer tourism took their place. Scrub and new forest appeared. In the 1920s and 1930s, deer were reintroduced, perhaps to give wealthy summer-home owners an illusion of untouched nature. Deer ticks returned and replaced local tick species. Thus arose the mixture of forest, wildlife, and urban escapees that made Lyme disease epidemic. It gave a new home to babesiosis as well.

Babesiosis is usually less severe than Lyme disease, but it can be chronic and dangerous, especially to the elderly and to people with impaired immune systems. So far, human infections have been relatively few, but it is now endemic on Martha's Vineyard and in parts of Long Island and Connecticut. Scattered cases have been reported across much of the Northeast and in Europe. Some people are infected with both Lyme disease and babesiosis. It remains to be seen whether babesiosis will continue to spread.

Just a year after epidemic Lyme disease, there came another new infection, more extensive and more dangerous. It arose from a quite different environmental change, the creation of controlled indoor climates with cooling and hot-water systems. Ironically, improved medical technology also played a role. The disease surfaced in July 1976, as Philadelphia hosted a celebration of the nation's bicentennial. Tourists and conventioneers thronged to the city, among them thousands of members of the American Legion. Many Legionnaires stayed at the Bellevue Stratford, a big, elegant old hotel on Broad Street, in the center of town. Soon many were suffering from fever, chills, aching muscles, and a dry cough. As their fevers rose, many became confused or delirious; soon dozens were hospitalized with acute pneumonia. Then, one by one, victims died. None of the usual pneumonia germs could be found; there was fear of a deadly new contagion. Visitors fled the hotel and the city.

As the number of sick and dead grew, the story of the Legionnaires' disease reached the top of the national news, and it stayed there for six months. The public was frightened, health officials nonplussed. At first, doctors suspected a new strain of swine flu that had been predicted for that summer, but flu tests were negative. They checked for pneumonic plague and psittacosis, or parrot fever; again results were negative. They eliminated the Lassa and Marburg viruses, then dozens of other micro-organisms. Finally they began to search for poisons and pollutants, including mercury, nickel, and pesticides.

In autumn, the epidemic was waning. Before it ended, 221 people had fallen sick, and 34 had died. The reason was no clearer than in July. Many victims were Legionnaires who had stayed at the Bellevue, but not all. The majority were middle-aged or elderly, many smoked, and some had chronic diseases such as diabetes. These were all risk factors for opportunistic pneumonia, but they did not explain the epidemic. Since the victims had not spread the disease to relatives and friends, it must have been transmitted by something in their shared environment, such as food, water, or air.

Food, however, was not to blame; some victims had never eaten at the Bellevue. Some had not even entered the hotel; they had only stood near it, on Broad Street. Epidemiologists tested and retested the hotel's water and air for microbes and toxins. One researcher guessed that germs from pigeon droppings on the roof had traveled by air to the corridors and rooms. Yet most guests and the hotel's employees had breathed the same air without getting sick.

Paranoia spawned wild theories. Frightened citizens and reporters asked whether terrorists had poisoned the Legionnaires. Perhaps a germ warfare agent had leaked from a train passing through Philadelphia. Perhaps the CIA was secretly testing a biological weapon on innocent civilians. The space exploration program might be introducing extrater-restrial viruses to earth. A researcher who had seen reports of penicillin-resistant gonorrhea bacilli wondered whether reveling Legionnaires had caught an exotic infection from prostitutes.

In January 1977, six months after the outbreak began, the CDC cultured a small, unfamiliar bacterium from the Legionnaires' blood. It was the sort bacteriologists call fastidious, able to survive in a laboratory only on special nutrients, within narrow ranges of temperature and acidity. Tests for such microbes were among the last and slowest carried out. This finicky germ was unlike any seen before. Over Legionnaires' objections, it was named *Legionella pneumophila*, and the disease was dubbed legionellosis.

As when any new infection appears, people asked whether it was new or just newly recognized. A study was made of blood samples hospitals had stored from unusual or unexplained pneumonias. Antibodies of *L. pneumophila* turned up in samples from a mysterious 1965 epidemic at St. Elizabeth Hospital, in Washington, D.C. It had sickened almost a hundred patients and staff and killed more than a dozen. The CDC linked *Legionella* to a 1957 epidemic in workers at an Austin, Texas, meat-packing plant, and to an isolated case from 1947. Legionnaires' disease was apparently some thirty years old in this country, and becoming increasingly common.

Surprisingly, *Legionella* antibodies also appeared in samples from a 1968 flulike epidemic in Pontiac, Michigan. There, 144 people had fallen sick but none severely or for long. Pontiac fever, as the illness was named, was a brief, mild legionellosis that struck healthy people of all ages, rather than favoring the aging and predisposed. Again the source was a building. All the Pontiac patients had in common was that they had been inside the headquarters of the state health department.

In 1977, as past epidemics were being identified, new ones broke out in hospitals in California, Vermont, and Nottingham, England; a quarter or more of the victims died. Over the following decade, more outbreaks occurred, from the United States to Holland to Australia. Everywhere they were linked to hotels, workplaces, and especially to hospitals. It was as if the buildings were toxic. Almost all were air-conditioned.

Air-conditioning was the clue that finally cracked the case in Philadelphia. Many of the Legionnaires said that the air conditioners in their hotel rooms wheezed or leaked. If *Legionella* could be spread by these machines, some puzzling aspects of the outbreak would be explained. Air-exhaust ducts might have infected the patients who had not entered the building but only stood near it. Hotel employees, with long, low-level exposure to the building's air, might have developed immunity without acute illness. But no other microbe lived and multiplied in air conditioners. How *Legionella* survived in such a biologically barren environment was a mystery.

Legionellosis is indeed spread by aerosol droplets. The usual sources are air conditioners, cooling towers, condensers, and hot-water systems. Some cases have been linked to whirlpool baths, humidifiers, and compressed air from steam turbines. Dust may have set off some outbreaks, such as the one at St. Elizabeth Hospital, where dirt from a nearby construction site apparently drifted in through open windows.

Legionellae are not exotic; they are common in nature. Since 1976, a dozen strains of L. pneumophila have been found; so have a dozen other species of Legionella. They normally exist in rivers, lakes, and perhaps soil, free-living or infecting protozoa. To cause human disease, they must be sprayed in the air, usually by manmade devices. That cannot have happened often before the 1950s, when air-conditioning became common. Legionellosis increased along with artificial indoor climate control and improvements in water and misting equipment. Medical science also helped the germ by making more people susceptible. Many victims have suppressed immune systems because of organ transplants or cancer therapy. Smokers, the elderly, and the chronically ill have been infected by respiratory equipment with which hospitals treated them.

Legionellosis can be severe, mild, or silent. An epidemic in New York City, traced to a cooling tower on the roof of Macy's department store, prompted a spot check of people working in the area; more than 1 in 10 had antibodies, reflecting past infection. The CDC records 500 to 1,000 cases each year, many serious or fatal, and estimates that the real number is 25,000 to 50,000. Legionellosis is now common in much of the world, and it is a major cause of hospital pneumonias.

Once the germ is established in a building, it can hang on for years. A possible explanation of its stubborn survival came in 1990, from a study of a three-year outbreak in a San Diego hospital that infected 26 patients and killed 10. The microbes were traced to the nozzles of shower heads, where they lived along with harmless amoebae. Such common, hardy protozoa offer nourishment to the bacteria. Chlorination and heat usually kill Legionella in water systems, but it is harder to get protozoa out and keep them out. Once they return, they are fodder for bacteria that then arrive.

Legionnaires' disease probably will keep increasing around the world along with air-conditioned interiors and hot-water systems. Technology also gives it new opportunities. As medical therapies that compromise immunity become more routine, legionellosis will have more victims. There are other sources; a 1990 outbreak in Bogalusa, Louisiana, killed 2 of the 34 people infected, and the germ was traced to a supermarket misting machine that moistened produce to make it more attractive.

Medical advances, ecological change, and altered behavior also helped create epidemic AIDS, the biggest and deadliest of the three new diseases. In fact, its emergence involved almost every possible source of a new infectious disease. In developed nations, the AIDS epidemic surfaced slowly and insidiously in the early 1980s, when doctors in New York and California noticed that some gay men were suffering from rare disorders. One was Kaposi's sarcoma, a skin cancer limited mostly to black Africans and some Mediterranean whites. Another was PCP, till then an extremely rare pneumonia caused by the protozoon *Pneumocystis carinii*. Yet another was toxoplasmosis, a protozoan brain infection. Eventually the patients' immune systems collapsed, and they died of these and other opportunistic diseases, such as tuberculosis.

At first, attention centered chiefly on Kaposi's sarcoma and patients' sexual orientation. Scientists tentatively called the syndrome GRID, for "gay-related immune disease"; sometimes the press referred to it as "gay cancer." By 1982, when AIDS received its present name, it had been found in Haiti, Europe, and Central Africa. In Africa it was spread mostly by heterosexual intercourse, especially prostitution. AIDS also had become a killer of hemophiliacs and intravenous drug users. With the dark humor that often arises when people cope daily with death, some health workers spoke of AIDS victims as the 4-H Club— homosexuals, Haitians, heroin users, and hemophiliacs. As the death toll mounted, and the range of the disease expanded, even dark humor became insupportable, and a growing trickle of heterosexually transmitted cases in developed nations began to influence perception of the disease.

By 1985, AIDS was pandemic. Cases had multiplied exponentially in the United States and Africa. By the early 1990s, it was epidemic in parts of Europe, Asia, and Latin America. Now the number of projected cases is staggering. Conservative estimates predict millions of infections in the United States early in the next century, and tens of millions worldwide. The number of deaths will keep chasing the number of infections, for AIDS is almost always fatal. Only slight progress has been made in easing and extending victims' lives, and no safe, effective vaccine is yet in view.

AIDS is realizing the nightmare visions provoked by such diseases as Ebola fever and hantavirus infections—a new pandemic that will cause not only individual tragedies but demographic disasters, leaving whole nations crippled as if a scythe had mowed down generations. The prospect evokes the Black Death, the smallpox die-offs of the age of exploration, and the worst years of the "white plague" of tuberculosis. The fact that the virus hides in the body for years before causing symptoms creates the specter of millions of silently infected people passing it on.

AIDS has aroused the sort of hate and repugnance that once greeted leprosy. Its deadliness is not the only reason; its frequent association with sex and drugs has elicited indifference or vengefulness toward its victims. They are often treated with more rage or fear than compassion. Initially, that fear was stoked by ignorance of the cause and transmission of the disease.

The original suspect list of pathogens included hepatitis B virus, Epstein-Barr virus, and cytomegalovirus. Swine fever virus was also high on the list; its symptoms in animals resemble those of people with AIDS. An AIDS outbreak in Belle Glade, Florida, a farm town west of Palm Beach, roused fear that mosquitoes carried the infection to people from pigs. The CDC was still studying that epidemic in 1983, when HIV, the human immunodeficiency virus, was discovered. Meanwhile, experiments with mosquitoes, pigs, and other species were showing that HIV occurs naturally only in humans. In Belle Glade, as elsewhere, AIDS was spread primarily by sex, intravenous drug use, transfusions, and other kinds of direct contact with infected blood and body fluids.

Inevitably, rumors of conspiracy and assault began to circulate, like medieval tales of plague-ridden corpses flung over city walls to poison the populace. Some of the tales have persisted. The disease was started by drug abuse. It was divine retribution for homosexuality and prostitution. Mutations caused by nuclear testing turned a harmless virus into a killer. Malaria experiments accidentally infected researchers with a virus from primates, and they spread it. HIV was engineered by the U.S. government to undermine Communist countries. It arose from genetic recombination in viruses in the monkey kidneys used to make polio vaccine for Third World nations. Many black Americans believe that HIV is a government invention devised to wipe them out, the drug AZT part of a plot to poison them, and condom education a genocidal ruse.

There is no good evidence for these ideas, and much against them. The more one knows about HIV, the less one can believe that anyone smart enough to engineer a virus would be dumb enough to try it with

HIV. It is one of nature's most unusual and, till recently, most baffling microbes.

HIV belongs to the family of retroviruses. By a remarkable stroke of timing, they were discovered in 1970 and first cultured in laboratories in 1978, just as the AIDS pandemic was starting. Retroviruses especially challenged researchers because of their structure and style of reproduction.

Many viruses, and virtually all other organisms, reproduce by duplicating their genetic DNA. This involves two major copying processes. First, DNA acts as a template to create reverse-image strands of RNA; it is helped to do this by the enzyme transcriptase. Then, the RNA in turn produces new DNA. In many long-term viral infections, from herpes (cold sores) to HIV/AIDS, the DNA may hide among the cell's genes and be transmitted to daughter cells. During its latency period, the virus cannot be killed without assaulting the host cell's nucleus.

In the 1960s, the duplication of DNA was thought to underlie all life. Then, many viruses were found to have genetic cores of RNA. This created novel ideas of life originating in a primitive "RNA world." It was largely a matter for specialists in virology and evolutionary biology. The RNA retroviruses were then discovered, and the diseases they caused were recognized. It was now urgent to understand how they reproduced.

After this type of virus invades a cell, its RNA produces DNA with help from an enzyme called reverse transcriptase—hence the name retrovirus. RNA viruses are structurally more prone to mutation than DNA viruses; retroviruses are spectacularly likely to mutate and create multiple new strains. It was clear to researchers from the start that developing drugs and vaccines against retroviruses would be a long, difficult task, like the first efforts a century earlier to kill bacteria without destroying their hosts.

There is nothing new about retroviruses. HIV-like viruses probably developed hundreds of thousands or millions of years ago in mammals. They all have an affinity for the T cells that play a crucial role in the immune system. Feline immunodeficiency virus (FIV) infects domestic cats, captive lions and tigers, and the Florida wild panther. Similar viruses live in horses, cattle, sheep, and goats. Fortunately, none of these can be transmitted to humans, and some are useful in studying AIDS. Much AIDS research is done on the related diseases that occur in simians (simian AIDS, or SAIDS) and in mice (murine AIDS, or MAIDS).

AIDS was the third retrovirus disease found in humans. The first, adult T-cell leukemia (ATL), was discovered in southern Japan in 1977. Like AIDS, it seems to be spread primarily by sex, contaminated needles, and blood transfusions. Half of those infected are male homosexuals and intravenous drug users. The virus can lie dormant for twenty or even forty years before triggering rapidly fatal leukemia (for which treatment is now becoming available). The cause, human T-cell lymphotropic virus, or HTLV 1, was discovered in 1980. It may have originated in Africa and traveled to the Caribbean and Japan in the days of the African slave trade. Today it occurs on every continent; it was recognized in the United States in the mid-1980s. There are speculations that HTLV 1 has an indirect role in some autoimmune diseases. If it does, the virus is more widespread, and has a much bigger impact, than was first believed.

In 1982, a close relative of that virus was found and labeled HTLV 2. It causes hairy-cell leukemia, so called for the appearance of the cells it affects. Like HTLV 1, it probably originated in Africa, traveled worldwide, and is spread in large part by drugs and needles. Today HTLV 1 and 2, like HIV, are usually screened from the donated blood supplies of developed nations, but they continue to reach people by other routes.

HIV, like HTLV 1 and 2, is a retrovirus. These viruses or very similar ancestors may have lived in African primates for countless millennia. We cannot pinpoint an epidemiological "big bang" when they first leaped to humans; such clear landmarks exist for few diseases. Usually a microbe in the environment or in another species causes isolated human infections; then environmental change, altered human behavior, or mutations in the germ make human epidemics possible. It is less an event than a process.

There are two ways to approximate when such viruses reached and adapted to humans. One is to screen old medical records and blood samples for cases misidentified in the past. The second, which became possible only recently, relies on new techniques of molecular biology. It creates an evolutionary tree linking a virus to its relatives; it determines how many genes they have in common, and how many mutations distinguish them. Assuming that mutations take a given amount of time to accumulate, one uses this evolutionary clock to calculate the emergence of one virus from another. Both types of research have pitfalls and uncertainties, but together they allow an educated guess.

Clinical records and blood samples have revealed a few possible cases of AIDS as early as the 1950s and early 1960s. It became more common in Africa in the 1960s, and in the United States in the early 1980s. Genetic studies suggest that the first type of HIV to be discovered, HIV 1, evolved in Central Africa; HIV 2 arose a little later in West Africa. Both are genetically related to simian immunodeficiency virus (SIV), which infects African monkeys without causing symptoms; apparently virus and host have had time to adapt to each other. SIV does kill Asian monkeys, to which it seems to be new. HIV 1, HIV 2, and SIV all probably evolved from a common ancestor in African monkeys, perhaps the green monkey and the sooty mangabey. Today's best evidence is that it happened no less than forty and no more than 250 years ago.

Despite the genetic calendars, which are tentative, some researchers believe HIV has existed in African primates for millions of years. Some even claim it has existed globally, causing sporadic cases and local mini-epidemics. Still others think it existed for centuries in isolated African tribes that came to tolerate it; the virus then spread to other peoples with no previous exposure, perhaps aided by mutations that made it more virulent.

Any of these theories may be true, but the most popular theory, and I think the most likely one, is that HIV reached humans some fifty years ago from African monkeys. The deadliness of AIDS in Africa suggests that it is as new to people there as elsewhere. It is difficult to imagine that a disease as distinctive as this would have long escaped the notice of native peoples and colonial health workers.* Still, debate over the origin of AIDS continues, sometimes tinged with anger and charges of racism by African government and health officials. AIDS is an export no one wants credit for.

The fight against AIDS is complicated by the tendency of RNA viruses to mutate often and create a number of types. Some people are infected with both HIV 1 and HIV 2, and with other types as well.

---

* This is a very brief summary of a complex issue. For details, and some different conclusions, one can consult such books as *History of AIDS*, by Mirko Grimek (see Bibliography for this chapter).

Several new varieties are being studied, and it may be only a matter of time before they all spread and mix worldwide. Furthermore, after HIV infects a person, it keeps mutating to evade that host's immune defenses. By the time symptoms appear, a person carries many slightly different varieties of the virus. This is one of the toughest hurdles in developing drugs and vaccines against it.

Since HIV and similar retroviruses are so widespread and common in nature, one must wonder why AIDS did not arrive sooner. One reason is that HIV is so difficult to transmit. It spreads by direct contact with blood or by repeated contact with such body fluids as semen and vaginal lubrication; it can also cross the placental barrier and infect a fetus. Although a single sexual contact can transmit HIV, it more often takes dozens or hundreds of exposures.

For such a virus to cross a species barrier and spread to tens of millions of people, it needs a lot of help. The help came from new technology, a degraded environment, and social and behavorial changes. Many of the changes were meant to enhance human health, prosperity, and opportunity.

HIV probably reached its first human hosts as most zoonoses have, from animals hunted for meat, kept as pets, or scavenging around human settlements. Many people in Africa keep monkeys as pets, and sometimes receive bites and scratches from them. That is probably how many retroviruses, such as FIV, are transmitted in nature. But in much of Africa, a monkey is less a friend than meat on the move. This was especially true as vaccination and insect control reduced infant mortality and sent populations soaring.

A rising population meant protein shortage, so people supplemented their diets with small game. They caught, killed, and skinned monkeys; sometimes they cooked them, and thereby killed most germs, but sometimes they ate them half-cooked or raw. That was how paleolithic hunters caught trichinosis from bears, African boys caught monkeypox from monkeys, and American hunters catch tularemia from rabbits. And it may be how some Africans caught HIV or its immediate ancestor.

Feeding a rising population required more than small game. Africans began to fell vast expanses of rain forest for farming. This put them at the ecological frontier where humans and other species have always exchanged pathogens. Clearing land meant contact with virus-infected monkeys. Crops and dwellings drew monkeys as scavengers. Similar results came from developing the timber, minerals, and oil that postcolonial nations needed for export. Countries without mineral resources pushed the production of cash crops and thus escalated land clearance, the disruption of ecosystems, and contact with unfamiliar germs.

In many nations, in the 1960s, live monkeys became a major export. They were needed by the tens of thousands each year for research and for manufacturing vaccines. In equatorial Africa, as in Asia and Latin America, villagers went out into forests to bring monkeys back alive. The monkey trade was perilous to both hunters and hunted. People were scratched and bitten. Monkeys were crowded in holding pens, cargo planes, and lab cages. Primates from all over the world were caged and shipped together, and exchanged pathogens. The result was a crucible of viruses, many capable of mutating and genetically recombining. The monkey trade carried Ebola-type viruses and SIV from African monkeys to Asian monkeys, and sometimes to humans. We will never know what part this played in the evolution and spread of HIV, but it may have been considerable.

While Africa exported HIV, it imported medical technology. There, as in much of the developing world, the hypodermic needle became a symbol of Western science and power. It dispensed such miracles as vaccines, antitoxins, antibiotics, and transfusions. The fine steel needle and glass syringe, perfected in the third quarter of the nineteenth century, were easy to sterilize; people who used them were routinely taught to do so. However, unsterilized hypodermics were often reused in Third World nations, and they spread malaria, hepatitis, and eventually AIDS.

Around the globe, hypodermics transmitted more infections than ever after World War II. Intravenous drug use increased, and with it the sharing of unsterilized needles. Some addicts sold their blood for drug money; if infected, they passed their diseases to the recipients. Others obtained drug money by prostitution and thus transmitted diseases they had caught through needles. A deadly synergy of addiction, prostitution, and other risky behaviors accelerated the AIDS pandemic. Such transmission became even more common after 1970, when cheap, disposable plastic syringes came into use. Hypodermic sterilization was no longer taught, but reuse continued.

Another major pathway for HIV is the sores of syphilis, chancroid, and other STDs. As we will see later, STDs have increased in the United States because of changes in attitudes and behavior. They soared in Africa when villagers poured into fast-growing cities to seek paying work. Rapid urbanization always brings exposure to crowd diseases and STDs. Many men who migrated to cities left behind wives, with the intention of sending money home or having their families join them. But where single men crowd together, liaisons and prostitution thrive. Besides, multiple mateships are part of traditional culture in parts of Africa; for many men there was no stigma on having sex partners in both city and village.

In changing Africa, a man who once would have spent his life in a village, with few potential sex partners, now might drive a truck back and forth across the continent, having sex with casual partners or prostitutes in several countries and dozens of towns. Women, left to cope on their own, many with children and no hope of support or paying work in villages, also migrated, and some turned to prostitution. As STDS multiplied, HIV was transmitted from men to women, from women to men, and from mothers to unborn children. The same changes now are spreading AIDS at frightening speed in parts of South America and Asia.

HIV thus traveled from villages to cities, and from cities back to villages. It did not help that many nations suffered tribal feuds, civil wars, and territorial conflicts. Famine and anarchy set millions of people adrift, carrying infections and susceptible to new ones.

International travel took HIV to North America, Haiti, and Europe and kept it circulating from continent to continent. Africans went to Asia, Europe, and the Americas for business, government affairs, and education. Developed nations sent to Africa business people, government and technical personnel, sometimes troops or mercenaries. Tourists traveled to view scenery and wildlife. Heterosexuals and homosexuals took erotic holidays to Africa one year, to Haiti, Thailand, or the Philippines the next, picking up and spreading a lethal cocktail of venereal germs. Airline and oil industry workers shuttled between Africa, Asia, and the Americas. The HIV equivalent of Typhoid Mary was the infamous index case, the first known carrier of the epidemic. This airline employee, knowing he was infected, had unprotected sex with hundreds of partners around the world.

Throughout history, social and technological change have ushered in deadly new epidemics, from bubonic plague to typhus. Today the great pandemic of change is AIDS. Like those of the past, it has created a demonology and an apocalyptic vision. Many of the elements that helped it emerge in Africa—ecological and population shifts, changed social and sexual patterns, technological innovations, increased travel—have also occurred in varying ways and degrees in other parts of the world. Globally, AIDS is not a uniform pandemic but many overlapping epidemics, with different viral strains spreading in different ways among various subpopulations. That makes the future seem in some ways hopeful, in other ways grim.

Research may not quickly produce cures or vaccines; probably drugs will first be devised to retard the disease process. Education has brought about some behavioral changes (condom use and fewer partners) and has reduced transmission, but behavioral control of disease is never universal and consistent. Needle sterilization has slowed HIV transmission in some populations, and much of the world's blood supply is screened for retroviruses. Furthermore, it is likely that the AIDS pandemic will slow after it has taken its worst toll among the most susceptible members of high-risk groups. The reduction of each AIDS subepidemic lessens the momentum of the larger epidemic. We can hope that before long, humans and HIV will evolve toward a better equilibrium, and the number of infected individuals will level off and then fall. The question is how many people will die before that happens.

Some prospects, however, are terrifying. Virologist Stephen Morse of Rockefeller University says that if any emerging virus could threaten the existence of our species, it may be HIV—especially if it mutates to an airborne disease, to become a sort of AIDS flu. According to Morse, there is no reason to think it could not happen. And if AIDS should, like the Black Death, kill 1 out of 3 or 4 humans, the resulting social breakdown could cause as much death and disaster as the infection.

Fortunately, there is as good a chance that mutations and human adaptation will make AIDS less lethal. Some infectious diseases, such as polio, become worse with time, but more often a host's immune system copes better with a pathogen over time. Some epidemic diseases take centuries to subside to chronic forms; others require only generations or decades. Syphilis changed from a florid, often lethal epidemic to a destructive but chronic infection in about two generations.

But by even the most optimistic scenario, AIDS will do horrendous damage. And even as we try to control it, or at least manage a holding action against it, we must face other questions. One is whether epidemics do to humans what the Machupo virus and plague bacillus do to rodents, make populations crash when they rise too high. It is a serious question as we watch the world's population threaten to double within the span of one human lifetime. Another is whether there is another AIDS, and yet another, waiting in the wings to emerge.

Twelve

inviting infection

Mad cows, dwarfs, and cannibals.
Infectious proteins. The slowest viruses:
contagious cancers. Sex and drugs,
drugs and sex. The white plague
returns. Toxic shock. Genes of
resistance and virulence.

In 1983, a humiliating malady killed George Balanchine, one of the century's greatest choreographers. Friends had seen him slowly lose his coordination and mental clarity. It was like watching a great painter go blind. His progressive paralysis and dementia mystified one specialist after another. Even after he died, physicians, dancers, and admirers continued to ask what malevolent illness had stolen his gift and then his life. On the first anniversary of his death, more than a dozen doctors gathered at Columbia University's College of Physicians and Surgeons to hear the postmortem findings and learn why they had failed to diagnose his sickness. His disease, though very rare, would soon arouse international interest and have implications for millions of patients.

The first sign that something was wrong came in 1978, when Balanchine began losing his balance while dancing. For a while, this was upstaged by a more urgent problem; he suffered from angina so severe that soon it confined him to bed with crushing chest pain. Coronary bypass surgery gave him relief in 1980, but by then his coordination was so poor that he had to talk dancers through routines instead of doing the steps with them. Despite repeated examinations, doctors could not explain why his equilibrium, eyesight, and hearing were deteriorating.

By 1982, it was obvious that Balanchine had a degenerative disease of the cerebellum, the part of the brain that governs balance and many other body functions. However, every test for a specific brain disease came out negative; only brain biopsy went untried, because Balanchine would not allow it. Later that year, he was hospitalized, and he lost his ability to walk, use his hands, and think clearly. His motor control dwindled until he even had trouble swallowing. One of the many frustrated doctors who attended him said, "We stood at the foot of the bed and shook our heads a lot. We thought he was dying of his own disease, one he invented." On April 30, 1983, he died of an unidentified malady.

When tissue from Balanchine's brain was examined under the microscope, it was spongy; there were so many spaces between cells that it looked like Swiss cheese. It also contained plaques of fibrous protein, the sort that appears with Alzheimer's disease. At last a diagnosis was possible. Balanchine had died of Creutzfeldt-Jakob disease (CJD), a condition so rare that it strikes just one person in a million. The only thing that would have revealed it was brain biopsy, the single test Balanchine had forbidden. Even if he had allowed a biopsy, his doctors could not have helped, for CJD is invariably fatal.

Although CJD was not recognized until this century, it belongs to a group of diseases whose history goes back at least 200 years, through such improbable victims as cannibals, dwarfs, mad cows, and stumbling sheep. In the eighteenth century, it was recorded in Iceland that a bizarre illness was killing flocks of sheep. First the animals staggered; then they trembled, became irritable, and itched so badly that they scraped off their wool on rocks and trees. In Britain, the relentless scratching gave the disease the name scrapie. For two centuries it sporadically decimated herds in Iceland, Scotland, and part of northern Europe, turning sheep's brains into spongy ruins.

In this century, scientists believed the disease was a viral infection, but they could not isolate its cause. Though apparently contagious, scrapie did not seem to pass from sheep to other species. It was puzzling that symptoms did not appear until several years after exposure to infection. The presumed virus was not only elusive and slow-acting but very hardy; it could withstand high temperatures and survive in the soil for years. In 1954, Icelandic virologist Björn Sigurdsson coined the term "slow virus" for the agents of scrapie and another sheep infection, visna, which produces something like human multiple sclerosis. Sigurdsson speculated that there were other viruses which worked silently in the body for years before causing symptoms.

The visna virus was discovered, but scrapie remained a murky byway of veterinary medicine. Then light came from an unexpected quarter. In the late 1950s, reports appeared of a strange new disease that was slaughtering the Fore tribe of Papua New Guinea. Virtually nothing was known of the disease, and little more of the Fore, who lived a Stone Age existence in New Guinea's isolated highlands. An adventurous young American virologist, D. Carleton Gajdusek, traveled to Fore territory to study the epidemic.

Nothing like this outbreak had ever been seen. The disease struck mostly women and children, causing loss of coordination, tremors, paralysis, and dementia; wasting and death followed the first symptoms within a year. The Fore called it *kuru*, the trembling sickness, and were convinced its cause was sorcery. Gajdusek had nothing to go on; the disease might be genetic, infectious, or the result of some environmental toxin. He began a one-man epidemiological study by walking from village to village for 2,000 miles, collecting data and tissue samples.

Gajdusek learned that some local outbreaks of kuru seemed to run in families, but there were enough scattered cases to indicate contagion rather than inheritance. He thought the most important clues to kuru must lie in its victims' brains. He begged mourners for autopsy rights and traded axes and tobacco for human brains. To his surprise, the microscope did not reveal the inflammation that usually accompanies viral brain infection. Instead he saw massive spongy degeneration and odd plaques of protein. This told him that although kuru seemed contagious, it had something in common with such noninfectious disorders as Alzheimer's, Parkinson's, and multiple sclerosis. Gajdusek immediately suspected that kuru had implications beyond New Guinea.

By the early 1960s, kuru had killed so many of the Fore that some villages held three men for every woman, orphans were ubiquitous, and the society was coming unglued. Gajdusek kept looking for a food, activity, or locale that linked the victims. Finally he discovered a task done only by women, along with the children in their care; they prepared Fore bodies for burial. These ministrations involved removing and handling the brains, and often women ate pieces of the brains. The custom of eating brains, they said, had entered Fore society only three or four decades earlier, perhaps from tribes to the north. That was about the time kuru first appeared.

Some scientists questioned Gajdusek's account of ritual cannibalism, more on the basis of theory than of observation. Gajdusek replies that even if these critics were right—and, he stresses, they are not—just handling infected brains could spread kuru through breaks in the skin. Because of their harsh life in the highlands, the Fore had scratches, cuts, and sores in abundance. The kuru agent entered their bodies by mouth, through their skin, or in both ways, and produced symptoms years or decades later.

Although kuru killed several thousand more Fore in the two decades after Gajdusek's discovery, the number of cases began to fall in the early 1960s, when the government strictly enforced its ban on cannibalism. Today kuru has almost vanished, but as Gajdusek suspected, its importance goes beyond New Guinea. The resemblance was soon noticed among kuru, CJD, and scrapie. All caused spongy degeneration of the brain, and all seemed to result from slow viruses. In the mid-1960s, Gajdusek injected brain matter from kuru victims into chimpanzees; within a year, they all died of spongy brain degeneration. For the first time, a degenerative disease of the nervous system had been transmitted. A year later, CJD was transmitted to lab animals the same way.

In 1976, Gajdusek won a Nobel Prize for his research, and the story of kuru and cannibalism briefly excited the news media. Then interest subsided, and slow-virus brain diseases received scant attention from both the public and researchers. The subject revived in 1986, when cows in Britain were dying of bovine spongy encephalopathy (BSE), better known as mad cow disease.

BSE resembled scrapie, CJD, and kuru. Cows began lurching and stumbling; as the disease progressed, they became irritable and aggressive, and then died. By 1994, despite vigorous efforts to control BSE, tens of thousands of cows had died or been slaughtered, and Britain could not export beef or even give it away; a donation of meat to chaotic post–Cold War Russia returned untouched. While no humans had caught BSE, fear of eating British meat persisted, for the cows had caught the disease by eating sheep.

This is not as unlikely as it sounds. In many countries, cow feed is enriched with some sort of protein; in Britain, one of the most common additives was sheep offal, including brains. Although scrapie had sporadically struck British sheep for at least a century or two, it had not been known to reach cows. That is apparently what happened in the 1980s. The question was why.

One theory has it that the 1970s oil crisis made feed producers turn to new, energy-efficient ways of rendering sheep carcasses; processing at lower temperatures may have let the scrapie agent survive. Another theory involves tallow, made of fat taken from sheep carcasses with powerful solvents. In 1980, the price of tallow fell, so renderers stopped extracting sheep fat. Perhaps the solvents had acted as germicides, and their absence after 1980 spared the scrapie agent. Either explanation or both may be true, and their timing is roughly correct. They put the transfer of disease from sheep to cows around 1980 or a little later; the scrapie agent usually has a latency period of three or more years.

Britain banned sheep offal from cattle feed, but BSE persisted; like scrapie, it can pass from pregnant females to their fetuses. Spongy encephalopathy also appeared in British domestic cats, American ranch mink, and captive elk and deer; all had eaten feed enriched with offal. Governments around the world continue to guard their borders against scrapie and BSE. Theoretically, both diseases could reach people through food, but apparently that has not happened. It may take eating other humans' brains, as the Fore did, to transmit such diseases among people.

It was not kuru, scrapie, or BSE that made most people aware of slow viruses, it was HIV. As one might guess, these viruses are all related. Lentiviruses, or slow viruses, probably evolved and diversified very long ago, as they adapted to different hosts. They are one of the three subfamilies of retroviruses; the other two are oncoviruses, which cause cancers, and spumiviruses, which do not cause human disease. Some slow viruses, such as HIV, attack primarily the immune system, and secondarily the brain. Others, such as the visna, kuru, and CJD agents, attack the brain.

So far, four spongy encephalopathies have been found in humans. Besides kuru and CJD, there is the rarer Gerstmann-Sträussler syndrome (GSS) and the still rarer fatal familial insomnia (FFI), discovered only in 1992. Small epidemics of CJD have struck Libyan Jews and Slovakian shepherds; there is debate about whether the reason is inheritance, infection, or both. CJD also killed dozens of children treated from the 1960s to the mid-1980s with pituitary growth hormone. Afflicted with dwarfism, the children had received hormone extracted from human cadavers, some of which must have carried CJD. Today dwarfism is treated with genetically engineered hormone, so that source of CJD has vanished, but some cases have more recently resulted from cornea transplants, brain-tissue grafts, and contaminated surgical instruments.

Molecular biology is beginning to explain how such diseases can be both inherited and infectious. CJD and its relatives result from alteration of a protein that occurs naturally in the brain. In some people, a genetic mutation makes the protein defective; in others, a virus affects cell metabolism and has the same result. For either reason, the altered protein accumulates in fibrous plaques and causes degeneration of several types of brain cells.

In 1982, Stanley Prusiner suggested that the cause of these diseases is not an ordinary virus but a protein molecule he calls a prion. The molecule, he says, invades the body and evades the immune system; despite its lack of genes, it manages to multiply, perhaps by hijacking the host cell's genes and acting as a template for copies of itself. This rogue protein mimics a virus and pretty much acts like one.

Some researchers accept Prusiner's theory of an infectious protein. Some think such molecules exist but must carry at least a little genetic material of their own. Others, including Gajdusek, believe a conventional slow virus will be found. In fact, Gajdusek thinks a kuru-type virus has been epidemic worldwide since the 1970s. Whether viruses or prions, such agents may hold the keys to several common and tragic disorders. For every Balanchine with a rare pre-senile dementia, there are millions who suffer from Alzheimer's, Parkinson's, multiple sclerosis, and other degenerative neurological diseases. For decades it has been suspected that viruses contribute to these illnesses, and perhaps to some cases of depression and schizophrenia. If these conditions are caused wholly or in part by slow viruses, researchers will have to start thinking of them as pandemics, and explain what changes in the organisms, the environment, or human behavior have made them so widespread.

Some of the slowest viruses are oncoviruses, which contribute to cancers; like HIV, they are becoming pandemic because of changing behavior and technology. As early as 1911, it was discovered that the Rous sarcoma virus causes tumors in birds, but decades of searching for microbes in human cancers ended in failure. Cancer research turned to genetics and the body's regulation of cell division. Oncogenes and oncoviruses were discovered in humans in the 1960s, and the two streams of research came together.

The first human oncovirus to be found was the Epstein-Barr virus (EBV). In developed nations, it usually infects adolescents, bringing mononucleosis. In poor nations, it usually infects children during the first year of life and, in the genetically predisposed, leads to Burkitt's lymphoma or to nasopharyngeal cancer. These two cancers are common in Africa but rare elsewhere, and some researchers suspect that malaria may be a cofactor in their development.

About 200 viruses are so far known to contribute to tumors in plants and animals, but cancer does not result from viruses alone. Its main or most immediate cause is disruption of the genes that regulate cell division. This happens in many steps over a long period, and usually for more than one reason. One cause is inherited genetic defects or mutations. Another is viruses that integrate their genes with those of the host. And many mutations and viruses produce cancer only if aided by such cofactors as chemical irritants, imbalanced diet, excessive sunlight, radiation, or other infections. Thus oncogenes, viruses, lifestyles, environmental insult, and other microbes can combine to cause cancer.

Viruses play a role in an estimated 15 to 20 percent of human cancers. We saw that the retroviruses HTLV 1 and 2, once rare and geographically limited, have become increasingly common around the world and can result in fatal leukemia. Suspicion has arisen that HIV may help induce cancer in some people. Two other viruses are creating cancer epidemics. One of them is hepatitis B virus (HBV), the major cause of liver cancer; it is among the world's most frequent and deadly malignancies. Its surge in the United States reflects a dismal failure in the fight against preventable infections.

There are five or possibly six kinds of hepatitis, caused by different and unrelated pathogens. HBV is an ancient human companion, probably inherited from our primate ancestors. Today it infects 200 million people, some 5 percent of the world's population, and kills more people than AIDS. The virus is ubiquitous in Southeast Asia, common in Africa, and increasingly widespread in Europe and North America. In the United States, there are 300,000 new cases of hepatitis B annually; 1.5 million Americans carry HBV as a result of past infections, for once in the body, it stays there for life. And wherever HBV is common, so is death by cirrhosis or cancer of the liver. Each year 5,000 Americans die of HBV-related cirrhosis, and 1,200 of liver cancer. The numbers have been rising for decades.

In developing countries, HBV usually reaches children from their mothers. In developed nations, it usually strikes adults. Like HIV, it is most often spread by blood, needles, and sex, but it is much more contagious. Although it can be mild, often it causes weeks or months of fever, jaundice, and fatigue. There is no reliable cure, and hepatitis B becomes chronic in one-tenth of the adults and the majority of children it strikes, progressively scarring the liver.

Half of the American victims of hepatitis B belong to one of five high-risk groups, each made vulnerable by its behavior. When HBV was discovered, in the 1960s, it was becoming one of the most common and deadly STDs of gay men. It was rampant and increasing among intravenous drug users. Recipients of blood transfusions were often victims. Doctors and other health professionals were picking up HBV from patients in growing numbers. And a new high-risk group emerged, heterosexuals with multiple sex partners.

A highly effective HBV vaccine appeared in 1982, the first vaccine devised against a cause of cancer. The results have been confounding. One would have expected susceptible people to rush to receive the series of three injections that give immunity. Instead of dwindling, the number of cases doubled in a decade. Hepatitis B is the only disease to have increased after the development of an effective vaccine.

As with AIDS, the hepatitis B epidemic consists of overlapping subepidemics. After 1982, safer sex practices slowed the spread of HBV in gay men in the United States, and screening virtually removed it from the nation's blood supply. Yet more drug users than ever caught it, and heterosexual transmission doubled. Most surprising was the failure of more than half of America's health professionals to seek vaccination.

Doctors' lack of knowledge and alarm about hepatitis B jeopardized not only them but their patients. A recent survey showed that 90 percent of doctors are unaware that heterosexuals with multiple partners are at high risk. Therefore they do not warn such patients against HBV or urge them to accept vaccination. Health organizations recommended in vain in 1991 that all children receive HBV vaccine, and in 1992 that all doctors do so. Hepatitis B has become the fourth most common reportable infection in the United States. Even if transmission were cut dramatically today, HBV-related cirrhosis and cancer would increase for decades, because so many people have already acquired this slow-acting virus.

Hepatitis C is creating similar problems. Like hepatitis B, it started to become epidemic around the 1960s. The virus was isolated only recently, and no fully reliable treatment or vaccine exists. It is spread the same way as HBV, and it has become even more common. It may kill more people than hepatitis B, some of them by cirrhosis and liver cancer.

The third newly epidemic cancer is cancer of the cervix. It, too, has increased lately because of changing behavior. The cause was discovered only after cervical cancer had been mistakenly linked to genital herpes, an STD that surged dramatically in the 1960s. Oral herpes, caused by herpes simplex virus 1 (HSV 1), produces a mild, flulike illness and cold sores, or fever blisters, of the lips and mouth. Its close relative HSV 2 results in a similar infection of the genitals. HSV 1 increasingly infects the genitals and HSV 2 the mouth, probably because oral sex with multiple partners has become more common.

Once HSV enters the body, it is there for life, hiding in nerve cells and waiting for new opportunities. Oral or genital symptoms may recur never, once or twice, or up to several times a month, when the virus is activated by fever, stress, fatigue, sunlight, or cold. The virus thus resembles another member of the herpes family, herpes zoster, which produces chicken pox in childhood and can be reactivated many decades later, resulting in shingles.

Although it was not rare in the past, genital herpes was not a wild epidemic until the 1960s. From 1966 to 1984, the number of cases rose fifteenfold. Herpes became a popular topic; patients formed support groups and discussed their physical pain and feelings of shame on television. There was increased concern about infants; herpes can be fatal for those who catch it from infected mothers as they pass through the birth canal or from other newborns in hospitals. It became journalistic wisdom that genital herpes would reverse the sexual revolution. Neither sex nor herpes vanished, but HSV was upstaged in the 1980s by AIDS. Today more than 30 million Americans carry HSV, and a half-million new cases appear each year. No cure exists, though now antiviral drugs can reduce the symptoms.

During the 1970s, few laymen knew it, but many scientists feared herpes might be lethal. Its rise had coincided with an upsurge of cervical cancer, and evidence was mounting that such malignancies are not only contagious but sexually transmitted. The earlier a woman begins coitus, the more partners she has, and the more often she has sex with them, the higher her odds of developing cervical cancer. A study of 13,000 nuns in the 1970s turned up virtually no cervical cancer.

These data suggested powerfully that a sexually transmitted virus was responsible. Some studies showed higher rates of cervical cancer in women with uncircumcised partners; perhaps the virus could hide under the foreskin. HSV was the leading suspect until 1977, when Harald zur Hausen, in Germany, showed that the agent of cervical cancer is not HSV but the even more common human papilloma virus (HPV), the cause of genital warts.

Genital warts, like hepatitis B and genital herpes, had probably been around for countless thousands of years, but they did not become rampant worldwide until the 1960s. The United States had 170,000 new cases in 1960, 2 million in 1986. From 25 to 40 million Americans now carry the virus, and in some cities it is found in more than half of sexually active teenagers. There are scores of strains of HPV, each with a predilection for a particular part of the body; at least four cause warts in the anal-genital region. These warts are usually harmless; they may vanish after two or three years, or they can be removed surgically. (Some stubborn cases of HPV and of hepatitis B and C are now being treated, with some success, with the messenger protein interferon alpha.) However, genital warts are contagious, and at least two strains of HPV can lead to cancer of the cervix, vagina, penis, or anus. Cervical cancer is now classed as an STD; women whose sex behavior puts them at high risk should have frequent pelvic examinations and Pap tests, to detect precancerous conditions while they are easily treatable.

The health effects of changed sex behavior, intravenous drug use, and new medical technology will take many decades to play out, especially in terms of slow viruses and cancers. We can expect rising rates of liver and cervical cancer, occurring at younger ages than in the past. Recently we have learned that bacteria as well as viruses can help cause cancer. The bacterium *Helicobacter pylori* was discovered a century ago; it was estimated to infect half the people in the world, but until the past decade it was not linked to any disease. It seemed remarkable only for its ability to thrive in the acid lining of the stomach. Now it is almost certain that chronic *H. pylori* infection can lead to ulcers and stomach cancer. The infection is more likely to do so in some people than in others; individuals with type O blood, for instance, are more disposed to give the germ a berth. Fortunately, antibiotics usually can wipe out *H. pylori* infections.

It is alarming, to say the least, that microbes and heedless behavior have made some cancers epidemic. However, this knowledge offers the possibility of preventing such cancers with vaccines and treating them with drugs, and of avoiding such cofactors as chemical carcinogens, excessive sunlight, and unprotected sex. And such epidemics as AIDS, hepatitis, and cervical cancer are stimulating research on the behaviors that launch and sustain them. This is crucial to control and prevention, since antibiotics and sexual and social behaviors are partly responsible for a shift in the whole spectrum of STDs.

When antibiotics first became available, in the 1940s, they were marvelously effective against syphilis and gonorrhea, then the most common STDs. One wonders whether the sexual revolution that followed would have had quite the same vigor if people had not believed that the venereal "wages of sin" were headed for insolvency. A half-century after the introduction of antibiotics, the results were not what anyone had imagined. New infections appeared, old ones became pandemics, and some dwindled only to return in hardier forms. The microbes adapted to new drugs and to new behaviors.

During the 1950s, syphilis decreased dramatically in the United States. Around 1960, it was returning in a series of new sub-epidemics. First it increased in gay men. In the 1980s, the practice of safer sex slowed that trend, but syphilis rose among poor urban blacks and Hispanics, chiefly because crack-cocaine use led to the trading of sex for drugs. By 1990, there were as many cases of syphilis as before 1950. Instead of vanishing, syphilis now infects 120,000 Americans each year.

Gonorrhea, once the most common STD, began to decline in the 1940s. It continues to do so, although there are still more than a million new cases each year. As gonorrhea declined, though, a similar STD more than filled the gap. Doctors had been puzzled for decades about the cause of nongonococcal urethritis (NGU), a gonorrhea-like disease that occurs without the presence of gonorrhea bacilli. In the past two decades, it has been learned that several germs can cause NGU; by far the most common is the tiny bacterium *Chlamydia trachomatis.* Chlamydia infection is not new, but as an STD pandemic it is.

In developing nations, *C. trachomatis* infects the conjunctiva, the delicate membrane lining the eyelids. Such trachoma may have afflicted the earliest humans; it probably became epidemic with the rise of village and city life. Trachoma existed in ancient Egypt and Greece, and it remains common where crowding, poverty, and poor hygiene allow its transmission by flies and dirty fingers. It is still the world's leading cause of preventable blindness.

In developed nations, trachoma of the eyes has almost vanished, but the microbe has survived by moving from the moist, sheltered lining of the eyelids to the moist, sheltered lining of the genitals. Chlamydia is now a pandemic STD of both sexes. It is the second most common infectious disease in the United States, after the common cold, with 4 to 6 million new cases a year. Often there are no symptoms, especially in women; the silently infected are chronic carriers.

Untreated, chlamydia infection can lead to pelvic inflammatory disease and scarred fallopian tubes. It is now the country's main cause not of preventable blindness but of preventable female sterility. Yet education about chlamydia remains so poor that many people learn about it only when they are diagnosed as having it. Like gonorrhea, it is becoming more difficult to treat because some strains of the germ are developing resistance to antibiotics.

Other STDs have increased as well. Chancroid, a bacterial infection that usually attacks men, declined in the United States because of penicillin and then returned. It has increased almost tenfold in the past decade. Giardiasis and amebiasis, caused by protozoa, and infections by salmonella and shigella bacteria, are all usually spread by contaminated water and food. However, they are more common as STDs; one reason is male homosexual practices, and another is that some people, thinking they are engaging in safer sex, now engage in oral-anal games.

The resurgence of so many STDs has made it easier to pass on and receive HIV. Diseases such as herpes, syphilis, and chancroid cause sores or minute breaks in the skin and mucous membranes that invite HIV directly into the bloodstream. Equally important, the body responds to STDs, as to most infections, with local inflammation, which attracts white blood cells. These cells are HIV's main hideout in the body. Their greater presence in the genitals in the course of any STD seems to raise the odds that HIV will be passed on to a partner.

The worldwide increase in STDs, and their shifting patterns and synergy, are not, as some moralists claim, a simple matter of the sexually greedy hastening to their deaths through sin and self-indulgence. STDs have different transmission patterns in different places and in various segments of society. In the Third World, sex behavior and STDs have changed because of exploding populations, altered land use, urbanization, migration, travel, technology, and the interaction of traditional and nontraditional cultures. In developed nations, some of the reasons are different. And in sex and STDs, as in so many things, there is one pattern for the prosperous and another for the poor.

Prosperous people in developed nations become physically mature and sexually active earlier than their parents, partly because of better nutrition and living conditions. They marry later, divorce more, live longer, and probably are sexually active till later in life. As a result, they spend far more years as single, sexually active adults—an invitation to STDs in all but the most cautious.

Among the poor in developed nations, better nutrition and health care have reduced infant mortality but not births. The result is a Third World "population bulge," a disproportionate increase of adolescents and young adults. These are the people most given to impulsive, risky behavior, from daredevil driving and drug abuse to early, unprotected sex.

Trying to control STDs by education and behavioral change has had mixed results. Disease transmission has slowed in gay men who practice safer sex and in drug users who take part in needle-exchange programs. But many AIDS and other STD programs repeat the mistakes of the futile anti-syphilis efforts of the World War I era. Moralizing and scare tactics, while they increase fear and guilt, may reduce risky sex behavior little or not at all; they drive many people away from testing and treatment, out of anxiety, defiance, or denial. Anti-STD programs may have their greatest effect on the "worried well" in low-risk groups, people who are already conscientious about health care. Many of the young, psychologically resistant, and poorly educated are not reached or do not listen. And many socially marginal individuals continue their high-risk behavior as if AIDS, syphilis, hepatitis, and cervical cancer did not exist. This hard core of the unknowing, heedless, or ineducable spreads new diseases and helps old ones evolve into lethal new forms. Nowhere is this clearer than in the recent emergence of multi-drug-resistant (MDR) tuberculosis.

Mycobacterium tuberculosis infects far more people than it makes sick. It is a slow-growing microbe with a waxy coat that protects it against immune defenses. Like a slow virus, it can lie dormant for years or decades; it manages to survive inside macrophages, the white blood cells that normally engulf and destroy invading bacteria. Anything that lowers disease resistance—poor nutrition, other infections, immune suppression—invites the bacillus to cause active disease. Once active, TB is highly contagious. In a crowded environment, one person can pass it by droplet transmission to dozens or even hundreds of others. In 1993, a flight attendant infected two dozen coworkers and a girl in a suburban California high school infected almost 400 other students.

The Industrial Revolution created ideal conditions for *M. tuberculosis,* through malnutrition, crowding, and poorly ventilated homes and factories. During the nineteenth century, as urban conditions improved, the spread of TB slowed, yet in 1900, the "white plague" still killed more people worldwide than any other infection. The 1930s brought the development of BCG vaccine; it gradually came into use in much of the world, though not in the United States. While the vaccine's effectiveness varies in adults, it protects millions of children each year from TB.

By midcentury, TB was second only to malaria as a deadly infection, but antibiotics could usually cure it. In 1953, the United States had 84,000 cases; in 1984, the figure hit an all-time low of 22,000. Doctors had come to think of TB as an atavism that occasionally reared its head in the poorest communities or among immigrants. A generation earlier, a bout with TB had been almost a *rite de passage* for young interns. Now TB, and TB research, had almost vanished.

Although the number of cases declined, the handful of remaining tuberculosis experts continued to warn that neglect could produce a renewal of epidemic TB. They also feared that the development of new drugs would not keep pace with the emergence of resistant strains of *M. tuberculosis.* No one listened, and funding for TB research and control almost ended in the 1970s. By 1985, when cases began to increase, many hospitals, old and new, lacked the ventilation systems, ultraviolet lights, isolation areas, and clinical expertise needed to treat active tuberculosis.

The disease returned for several reasons. One was the vulnerability of the immunosuppressed; TB is often the first sign that a person harbors HIV. Another was inadequate health care for the poor and socially marginal. At first, TB increased most visibly among drug addicts, prison inmates, and the homeless. Many of these people had little access to health care, and some refused care when it was offered. Worst of all, many quit treatment before it was complete. That was what gave the greatest boost to MDR microbes.

In 1984, half of the people with active tuberculosis had a strain of the germ that resisted at least one antibiotic. Today many strains resist four to seven drugs for treating TB, and some resist them all. This makes the disease more lethal, and treatment far longer and costlier—six to twenty-four months of therapy, costing tens or hundreds of thousands of dollars. Treating MDR tuberculosis is thus more difficult, and the odds are higher that people will quit treatment before it is complete.

One obstacle is that many patients start feeling better after just a few weeks of treatment. At that point, drugs have killed the most susceptible bacilli, but the germs genetically endowed with high resistance survive and multiply. The patient who prematurely quits treatment spreads these resistant bacteria. Developed nations have fewer airless slums to aid transmission than in the past, but *M. tuberculosis* travels efficiently in shelters for the homeless, school buses, and perhaps buildings and airplanes that use recirculated air. In these and other public places, MDR strains pass from high-risk to low-risk populations.

When people who have dropped out of treatment relapse and are treated again, they no longer respond to the drugs they first received. Some TB patients are hospitalized and check themselves out many times; each time they check out, they spread a larger number of resistant strains. Some of the most recalcitrant dropouts are drug-addicted, alcoholic, or mentally ill. Many have no permanent home and cannot be traced by routine public health procedures. In the early 1990s, when tuberculosis was clearly out of control, such people aroused fearsome memories of Typhoid Mary, who infected fifty people before being forcibly committed to a hospital in 1915. In some cities, aggressive outreach programs were started, to help people initiate and continue treatment for TB. The most stubbornly resistant patients were hospitalized against their will, something the AIDS epidemic had not provoked.

In 1987, when the resurgence of tuberculosis was new, the CDC predicted that with $36 million a year the United States could wipe out the disease by the year 2010. A recent projection calls for fifteen times as much money and is less optimistic. Tuberculosis has increased in New York City, Houston, and Miami, and the number of cases is rising in urban, suburban, and rural areas across the country. For the sick to be treated, and the epidemic contained, a new generation of TB researchers and clinicians must be trained. There is no quick fix.

Tuberculosis is even more threatening elsewhere in the world. In the years 1985–1991, when it increased 12 percent in the United States and 30 percent in Europe, it rose 300 percent in the parts of Africa where TB and HIV are inseparable. It is skyrocketing in Asia along with AIDS. The bacillus infects 10 to 15 million Americans and 1.7 billion people globally, about one-third of the human race. Approximately 10 million people have active tuberculosis; it kills 3 million each year, 9,000 every day. The numbers dwarf those for AIDS and malaria.

How many *M. tuberculosis* carriers develop active disease will depend on their standard of living and their general health. How many of the sick survive will depend on their medical care and whether drug research can outpace the evolution of new MDR bacilli. The problem of microbial drug resistance is approaching critical levels, and tuberculosis is only the biggest example. Sometimes the evolution of new strains creates epidemics of what are, in effect, new diseases.

Staphylococcus aureus is a common bacterium that causes boils and wound infections. In severe cases, its powerful toxin can be fatal.

Penicillin was first used on *S. aureus* in 1941, and it was effective. By 1944, some strains began showing resistance. Hospitals and institutions for the very young and very old, where antibiotics were used heavily, became breeding grounds for new, resistant strains. In the 1950s, a type of *S. aureus* appeared that caused serious infections in nurseries and pediatric wards around the world. Post-penicillin antibiotics more or less controlled it. Then new strains of "hospital staph" became even more widespread in the 1970s, especially in people treated with invasive procedures such as intravenous lines, dialysis catheters, and heart-valve implants. Antibiotics were developed to control these new strains, and the microbes became resistant to them in turn.

In 1980, hundreds of American women were struck by a powerful new infection that became known as toxic shock syndrome, or TSS. *S. aureus* was cultured from many victims' vaginas, and it was discovered that the infections had been triggered by a new, superabsorbent menstrual tampon that gave *S. aureus* an ideal environment for rapid growth. The tampon was removed from the market, but by the time the epidemic subsided, in 1982, more than 800 women had suffered TSS, and about 40 had died. The question remained whether TSS represented a new strain of *S. aureus* that produced a particularly potent toxin.

A similar epidemic seized public attention a decade later, through the death of a celebrity. In 1990, Muppet creator Jim Henson died suddenly of toxic-shock-like syndrome (TSLS), an infection that kills rapidly by pneumonia and shock. It resembles TSS but is caused by a group A streptococcus. In pre-antibiotic days, strep A resulted in scarlet fever, rheumatic fever, and wound infections. It produces a toxin that kills as easily as that of *S. aureus,* but more quickly. Doctors said that if Henson had sought treatment even a few hours earlier than he did, he would have survived.

Between the world wars, strep A surprised doctors by becoming less common and less severe, for no apparent reason. Strep throat became a usually nonlethal disease of children, treatable by antibiotics. In the mid-1980s, acute strep A infection reappeared in adults in Europe and the United States, with a death rate of 20 percent. As with TSS, the question is still whether an old virulent strain returned or a new one evolved. Whatever the case, doctors see TSS and TSLS as clinically new and increasingly common; both kill by whipping the immune system into a frenzied overreaction. And these are only two of many new varieties of microbes with high virulence that challenge existing treatments.

Genetics and molecular biology have shown that drug resistance and virulence do not change only through simple evolutionary selection, in the way described for *M. tuberculosis*. Bacteria can transmit genes to each other by exchanging chromosomal DNA or small bundles of genetic material called plasmids. Resistance and virulence genes can thus spread not only within a species of bacteria but from one species to another. Bacteriophages, the viruses that infect bacteria, can also transmit these genes as they move from one host to another.

Genetic exchanges can affect the kind and amount of toxin bacteria produce. They can likewise affect a microbe's ability to colonize certain types of tissue, such as the brain or the lungs. They can alter how a germ responds to a host's body temperature or to the amount of iron or calcium in the host's body. Hundreds of bacterial toxins and dozens of virulence genes have been discovered, and many more will be found. They help determine the kind, extent, and severity of illnesses bacteria cause in people. They offer medical science new avenues for dealing with old and new infections.

As these and other factors are studied, drugs continue to lose ground in the battle against tuberculosis, gonorrhea, chancroid, typhoid, cholera, meningitis, urinary and surgical infections, staph and strep infections, and diarrheal diseases that can kill small children. Drug resistance also appeared in the protozoon responsible for malaria and in some viruses.

For a half-century, we have tended to assume that if microbes become resistant to one antibiotic, another will soon arrive and save the day. But we may be running out of new antibiotics. To keep up with the speedy evolution of microbes, we probably will have to devise some radically different kinds of drugs—and we should expect microbes to adapt to them as well.

We are responsible for hastening the evolution of microbial resistance. Antibiotics have often been overprescribed in the United States, and they are used even more indiscriminately in many other countries. They have been fed in huge quantities to livestock, not to cure disease but to aid growth; this is a major reason for heightened microbial resistance. Not least important, hospitals, where antibiotics are used heavily, and where invasive and immunosuppressive therapies keep increasing, have helped create bacteria that shrug off many or all anti-biotics. About 1 hospital patient in 20 in this country acquires an infection before being discharged; that means some 4 million hospital infections a year, with many thousands of deaths.

There is no stepping back from the use of antibiotics, with all their benefits and their problems. We will continue to subject microbes to massive selective pressures, making some of them drug-resistant and affecting their virulence. To survive our own ingenuity, we will have to use drugs more prudently and set up a worldwide surveillance system of drug resistance. The alternative could be a return to the hospitals of the nineteenth century, with crowded infectious-disease wards full of incur-able patients. It was antisepsis and antibiotics that made hospitals places where people went to have their lives saved; it would be criminal to again make them places where many go expecting to die.

Thirteen

from this
time on

Vibrios resurgent. Fragile barriers.
Crowds, chaos, war, and pestilence.
Six ways to make a plague.
Partners in illness. A savage test. Hopes.

In 1960, the Americas had been free of cholera for so long that most doctors in this hemisphere had never seen it. The United States had had no major outbreak since 1866, South America since 1895. On other continents, the sixth pandemic had come and gone, but cholera remained endemic in its original home, Bangladesh and India, and in other parts of Asia where *Vibrio cholerae* reservoirs had developed in local waters.

In 1961, the seventh pandemic erupted, in Indonesia. Its source was the new El Tor strain of *V. cholerae*, hardier and more easily transmitted than its predecessor. El Tor spread through Asia and hit Africa in 1970.

There it attacked twenty-nine countries in two years; it remained sporadically active, with death rates as high as 10 percent.

In 1991, after thirty years, the El Tor outbreak was the longest
cholera pandemic ever. When it reached Peru, in January of that year,
the only surprise was that some dangerous contagion had not struck
sooner. Peru's headlong urbanization had been accompanied by a long
civil war that eroded the economy and infrastructure. Lima's population
had risen from 1 to 7 million in a few decades, and vast numbers of
people lived in crushing poverty. The city was surrounded with squalid
shantytowns; almost half the inhabitants lacked clean water and sewage
disposal. Inspired by an erroneous report that chlorination posed a
serious cancer risk, Peru's government had stopped chlorinating its
water.

El Tor apparently reached Peru in water flushed from the ballast
tanks of ships from Asia—the same way vibrios briefly infected Mobile
Bay, in the Gulf of Mexico, that same year. The Lima epidemic began
after people ate ceviche made of tainted raw fish from local waters. The
wastes of infected people entered Lima's antiquated sewers, drainage
ditches, and water systems. By late April, there had been 150,000 cases
of cholera, sometimes as many as 4,500 a day.

Since the 1970s, it has been possible to reduce cholera deaths
greatly by immediately replacing victims' lost fluids and body salts.
Thanks to this fast, cheap treatment, only 1 percent of Lima's cholera
victims died. Still, that meant 1,500 deaths in Peru in a few months.
Fear was rampant that cholera would race through slum-ringed cities
and poor villages all over Latin America.

Despite attempts to contain the epidemic, that is what happened.
Cholera reached a new country almost every month during 1991,
flashing through Colombia, Chile, Ecuador, Bolivia, Brazil, Argentina,
and Guatemala. By early 1992, it had sickened 400,000 Latin Americans, killed 4,000, and was approaching Mexico's border with the
United States.

This country was spared more than scattered cases and small,
isolated outbreaks; most involved imported foodstuffs or travelers returning from nations with epidemics. Still, there were long-range perils.
If V. cholerae infected plankton off the coasts of the Americas, they
would remain there after the epidemic ended, to cause new outbreaks
decades or centuries later. And if concern about global warming was

justified, warmer waters might extend the disease's reservoir farther north.

In 1993, before such fears could be realized or dismissed, a new strain of *V. cholerae* had erupted in India and Bangladesh and killed 5,000 people. Strain 0139 (the 139th to be discovered) soon spread to Southeast Asia; still on the move, it may signal the eighth pandemic. It can kill within hours and is even hardier than El Tor, which it may replace as it travels. Meanwhile, El Tor remains endemic in Latin America, where it has already stricken more than 2 million people.

Cholera's resurgence shows the fragility of our barriers against epidemics. Economic collapse, social disorder, war, famine, flood, or earthquake can easily leave great masses of people helpless before microbes. Even during brief disruptions, water and wastes go untreated, disease-carrying insects multiply uncontrolled, garbage brings scavengers and vermin, children are not immunized, and malnutrition helps sicken the healthy and kill the sick.

This happens somewhere in the world every year. Recently typhus, typhoid, yellow fever, malaria, dysentery, cholera, hepatitis, pneumonia, and other ills old and new have struck in starving Ethiopia, in war-torn Somalia and Rwanda and Afghanistan, and in chaotic former republics of the Soviet Union. It is not only in developing nations that epidemics rise from social and natural disruptions. In 1994, an earthquake in California stirred fungal spores in the soil and caused an outbreak of valley fever (airborne coccidioidomycosis). In the previous year, as we have seen, drought followed by heavy rains provoked a hantavirus epidemic in the Four Corners region of the Southwest.

Such events show why scientists often focus less on specific germs than on the ways microbes spread and reach epidemic intensity. In 1992, when a panel of distinguished researchers (chosen by the National Institute of Medicine) produced a landmark study of emerging infections in the United States, they concentrated on six major causes of disease emergence.

- Breakdown of public health measures
- Economic development and land use
- International travel and trade
- Technology and industry

- Human demographics and behavior
- Microbial adaptation and change

The potential for breakdowns of hygiene and public health measures was always obvious in poor countries; in 1992, the surprise was that they were happening also in wealthy nations. In the United States, flaws in the health care system, a growing underclass, and risky behavior were bringing back tuberculosis and spreading AIDS, hepatitis, and other diseases. Failure to maintain vaccination programs has resulted in the return of measles (a major epidemic in 1989–1991) and whooping cough (in 1993, more cases than in any year since 1967). In the former Soviet Union, the situation now suggests a slide back toward Third World conditions. Cholera, typhoid, hepatitis, diphtheria, dysentery, tuberculosis, and STDs are all increasing explosively. Among the reasons are poor hygiene, reuse of hypodermics, mass movements of sick refugees, failure to vaccinate, and environmental contamination.

The 1992 report failed to foresee one enormous new threat, that the end of the Cold War would speed destabilization in many parts of Europe, Asia, the Near East, Africa, and Latin America. Decades of social and political upheaval probably lie ahead. They will bring intense nationalistic struggles, social chaos, brushfire wars, and major armed clashes. From Yugoslavia to Yemen, civil wars have been followed by outbreaks of disease. In 1994 in Zaire, a million political refugees from neighboring Rwanda died by the tens of thousands, first from cholera, then from antibiotic-resistant shigella dysentery. Cases of measles, typhus, and bubonic plague followed, and perhaps pneumonia and meningitis; in fields littered with the dead and dying, more time went to hasty burial than to counting bodies and refining diagnoses. It was recently estimated that almost 20 million people around the world are now refugees, many living in dire health conditions.

One of the first signs of a failing public health system is the spread of waterborne and foodborne infections. These were long thought to lie mostly in the West's past, but they are epidemic everywhere, partly because of new and newly recognized diseases. In many poor nations, children suffer ten to twenty bouts of microbial diarrhea by age three; combined with malnutrition, it kills 4 to 10 million of them. In the

United States, typhoid and cholera remain rare, but other water- and foodborne diseases sicken millions and kill thousands each year.

As in Neolithic times, some of these infections are constantly reintroduced to humans from domesticated and wild animals, directly or indirectly, as when rains wash their wastes into sources of drinking water. Rotavirus, the world's leading cause of childhood diarrhea, seems to have originated in swine; Norwalk virus, the chief cause of adult diarrhea, also may reach humans from livestock. These viruses were discovered only in the 1970s; they probably became more common wherever human and herd populations exploded. Both are transmitted by water and then by unwashed hands, food handling, and swimming pools. They take high tolls among the very young and the very old, especially in the nurseries and old-age homes where more and more Americans spend their days.

Also increasingly widespread is the intestinal protozoon *Giardia lamblia*. Though described three centuries ago, it was proved to cause human disease just three decades ago. In the 1970s, there were major giardiasis outbreaks in St. Petersburg (then Leningrad), in upstate New York, and in Aspen and Vail, Colorado. Now *G. lamblia* is known to cause epidemics all over the world. Giardiasis is a typical zoonosis, spread by dog or beaver feces. Beaver, like deer, were almost wiped out in the American Northeast, but then thrived again. In 1986, beaver droppings in the backup reservoir of Springfield, Massachussetts, caused 7,000 cases of giardiasis. *G. lamblia* is also sexually transmitted among gay men. It may be responsible for many misdiagnosed cases of irritable bowel syndrome.

The bacterium *Campylobacter jejuni* and the protozoon *Cryptosporidium,* which cause major diarrhea epidemics, both reach humans from cattle. Cryptosporidia were not known to sicken humans until 1976; since then, they have set off big outbreaks of disease the world over. In 1993 in Milwaukee, 400,000 people fell sick because rainwater had washed cryptosporidia from dairy-herd wastes into the water supply. This prompted a national survey of water systems, many of which were found to have inadequate inspection and filtration. Fine filtration, not just chlorination, is needed to kill cryptosporidia and some other diarrheal microbes, yet 50 million Americans drink unfiltered water. Not surprisingly, waterborne germs continue to cause ills that result in days of discomfort, months of pain, and even death.

Thousands of people die in the United States each year because of infections by old, familiar foodborne bacteria such as salmonella and shigella—from eggs, improperly processed chicken, undercooked beef, and tainted seafood. These bacteria affect more people than a decade or so ago, because of automated, speeded-up methods of slaughter and processing in an increasingly consolidated food industry. About 40,000 cases of salmonellosis are reported each year, but the CDC estimates that there are actually between 1 and 4 million, thanks to a food inspection system inadequately funded and using methods eighty years old.

New foodborne diseases are appearing, some of them deadly. Until recently, *Listeria* bacteria rarely affected humans; now, transmitted by meat, chicken, and soft cheeses, they sicken and kill people in Europe and the United States. So does a new strain (0157:H7) of the common microbe *Escherichia coli,* transmitted to people from cattle. It can cause serious, even deadly, kidney damage, especially in children. Epidemics from Washington State to New Jersey have struck people who ate undercooked hamburger at home or in fast-food chains. This microbe turned lethal when it picked up from other bacteria a gene for producing a shigella-type toxin—probably in the process of developing resistance to the antibiotics fed to livestock.

Another dangerous new foodborne pathogen is the helminth anisakis, commonly known as the herring worm or codworm. As more people around the world have taken to eating uncooked or lightly cooked fish, anisakiasis has spread from Japan to Holland, the United States, and other countries. Efforts to keep our food free of this and other foodborne pathogens have made only a small dent. Health is being held hostage by corporate resistance, government budget battles, and ignorance.

The second major source of new diseases, according to the 1992 panel, is economic development and changing land use. Many examples have already been mentioned. Cutting down forests for farmland invited hemorrhagic fevers in South America and Africa. In Asia, irrigation and wet farming bred mosquitoes that carried malaria parasites and viruses. New dams brought Rift Valley fever to much of Africa. In the United States, reforestation and suburban sprawl spread Lyme disease, babesiosis, and Rocky Mountain spotted fever. Everywhere, urbanization, development, and industrialization alter landscapes, and disease patterns are transformed.

Changing land use and development doubtless will continue to create and disseminate infections in prosperous countries, but the results will be worse in poor ones. Rising populations create pressure for subsistence and for exports such as cash crops, oil, minerals, and timber. As cities and industry develop at breakneck pace, infrastructures and health measures are sure to lag. Such changes have always aided the appearance of new diseases and a rise of those spread by crowds, insects, pests, water, and food.

Of all the threats of development, the most serious may be global warming, caused chiefly by the burning of fossil fuels and massive deforestation. Some scientists think the recent warming trend is just a blip in natural climate cycles; others consider it a figment of erroneous computer models or environmentalist hysteria. At the other extreme are people convinced that carbon dioxide, methane, and other gases are turning the world into a foul greenhouse where our children will choke or drown.

The positions on both extreme are arguable. The world did grow somewhat warmer in the period 1880–1940, and again in recent decades. However, temperatures dropped enough in 1940–1960 to provoke predictions of an impending ice age. It now seems to be the majority view that some degree of warming is under way or soon will be. The major debate is between those who see the warming as slight and those who forecast a catastrophe.

The most dire predictions are numbing. They hold that before a century has passed, coastal cities and deltas will start sinking under saltwater. Stricken crops will set off famines. The despoiling of tropical forests will raise temperatures and shrink biological diversity in ways that threaten the entire biosphere. Species from butterflies and slugs to tigers and polar bears will become extinct. Greater heat and humidity will invite malaria, dengue, cholera, and insect-borne viruses into temperate zones. Obscure tropical parasites will become mass killers.

Before succumbing to panic, one should recall that climatic forecasts have a notoriously poor record. Like ecosystems, climate systems are complex beyond our power to predict. Should global warming increase, results both good and ill may be unlike anything we imagine. Certainly the number and kinds of diseases that would arise and spread remain informed guesswork. Sensible caution suggests that we consider the threat seriously and take basic steps against it. Many countries and international organizations are moving in that direction, but economic pressures may make it impossible for poor nations to live up to their intentions—another reason why those who can act should do so.

Next on the panel's list of sources of new diseases was travel and trade, which have caused and spread infections since nomads first ventured beyond their home territories. At first, new diseases could spread only as far and as fast as people could walk. Then they went as far and fast as horses could gallop and ships could sail. Wars and migrations have carried infections to virgin populations for at least several thousand years, and ships have probably broadcast epidemics at least since the plague of Athens. In the five centuries that travel has been truly global, most new diseases have been potential pandemics.

Ships continue to carry epidemics, through sick crewmen and passengers, infected ship rats, and mosquito eggs in damp cargo holds. At the turn of this century, ships brought the third bubonic plague pandemic from Asia to California, where it set off an outbreak in San Francisco's Chinatown. Then the bacillus literally went underground, becoming endemic in ground squirrels and prairie dogs. It is still there, creeping eastward, occasionally taking human lives when people or their pet cats come in contact with these animals.

Containerized cargo has reduced some of the risks posed by shipping, but the flushing of bilge by freighters, cruise ships, and supertankers continues to spread cholera and perhaps other diseases. In fact, taking on and releasing bilge can transport much of a marine ecosystem from one part of the world to another, with unpredictable consequences. One nasty result, with severe economic costs, has been the transport of zebra mussels from Asia to North American rivers and lakes, where they are overwhelming their new environment.

The worst risk of transporting infections now comes from airplanes. They enable tourists, business people, soldiers, migrant workers, and political refugees to make in hours trips that used to take weeks or months. People who are infected, especially those who are still asymptomatic, can thus spread HIV, tuberculosis, influenza, Lassa fever, and many new viruses. The airborne trade in lab animals and pet species is especially dangerous, as was proven by the 1989 Reston outbreak of an Ebola-type virus in lab primates.

Equally disturbing is the fact that often we do not know just how some microbes speed around the globe. In recent decades, several new types of cat and dog infections (feline and canine parvoviruses) have spread worldwide within a few years. Some of these viruses may have crossed from one species to others, though none has yet infected humans. In the 1990s, several other microbes have crossed species barriers and killed large numbers of dolphins, seals, and African big cats; these outbreaks may be linked to pollution and other manmade environmental changes. They pose a warning about how new human infections could arise and become ubiquitous in the future.

The only imaginable brake on the jet-borne spread of diseases is a rigorous international quarantine system, combined with extensive, expensive, and involuntary medical testing. Such measures could never be efficient and consistent enough to have much effect. Apparently we have no choice but to accept that now, as several times in our species' history, we will have to endure a painful adaptation to a bigger, more diverse disease pool.

Technology and industry was the next item on the panel's list. New technology appears at stunning speed, constantly opening new niches for microbes. Air-conditioning and other cooling and heating systems have made legionellosis common around the world. In similar fashion, highly mechanized food processing has spread diarrheal diseases. Perhaps most alarming is the growing number of new infections created by medical technology.

Medical inventiveness has vastly increased the ability to treat such conditions as infections, cardiovascular disease, and cancer. In doing so, it has raised the number of invasive procedures that invite new diseases. Every needle, catheter, probe, and intravenous line invites microbes into the body. Hospitals where such instruments are used breed *Legionella*, antibiotic-resistant staphylococci, and opportunistic fungi and viruses. Hospital infections occur by the millions each year in the United States, and far more often in poor countries.

Some medical technologies pose especially high risks. Donated blood and transplanted tissues have transmitted HIV and HTLV diseases, hepatitis B and C, and Creutzfeldt-Jakob disease. In developing nations, malaria and other diseases are commonly spread by blood supplies. Developed nations have greatly reduced transfusion and transplant infections, but a new crop of diseases has sprung up because of immune suppression created by HIV or by medical therapy itself.

A weakened immune system leaves the body prey to microbes it would otherwise fight off, and to some that normally dwell in it harmlessly. That is why AIDS patients suffer an array of opportunistic infections by TB bacteria, herpes virus and cytomegalovirus, *Cryptococcus* and *Candida* fungi, and such protozoa as *Giardia, Toxoplasma,* and *Pneumocystis carinii.* Some of these illnesses, such as *P. carinii* pneumonia and toxoplasmosis of the central nervous system, were almost unknown until AIDS became epidemic. AIDS patients are not the only victims of such infections. Organ transplants and treatments for cancer and autoimmune diseases often involve drug-induced immunosuppression; the patients are susceptible to many of the same ills that strike AIDS victims.

Improved medical technology is also saving premature infants and extending the lives of the elderly. The very young, with immature immune systems, and the elderly, with weakening immune defenses, are naturally somewhat immunocompromised. Infants in hospitals, small children in day care, and the old in group homes are therefore especially vulnerable to the infections common in institutional settings and induced by medical technology. This will be a bigger problem in the near future; according to some estimates, up to 20 percent of Americans will be over age sixty-five by the year 2030.

The fifth factor the 1992 panel listed as favoring disease emergence was changes in demographics and behavior. At the end of World War II, the world had about 2.5 billion people; now it has more than 5 billion. Even if steps are taken to slow population growth, it could double or even quadruple by the year 2050. The growth has leveled off in most wealthy nations, though there is no guarantee that this will not change. Elsewhere, growth remains explosive. In 1992, The Royal Society of London and the U.S. National Academy of Sciences made a rare joint statement, warning that if present population growth forecasts are correct, "science and technology will not be able to prevent either irreversible degradation of the environment or continued poverty for much of the world." Compared with some other predictions, this one is mild.

Opinions about whether the earth can feed so many people and tolerate their effluvia range from measured confidence to despair, but population models have as bad a predictive record as those for climates. If the answer is no, the choices are simple: We must stop population growth or die, and the draconian Chinese model for doing so shows cruel wisdom. But let us hopefully assume that humanity can manage to produce enough food for its rising ranks, and not expire amid wastes and pollutants. Many agriculture experts think this is possible; they point out that most famines result more from poverty and poor distribution than from shortage of food, and that productivity can be increased. But beyond sheer numbers, there remain the problems of urbanization and changing demographics.

Two centuries ago, 98 percent of the world's people were farmers and villagers. Soon half will be urbanites, many living in megacities of 10 million or more. Such cities put huge strains on water and waste systems, infrastructure, social order, and public health programs. At the same time, demographic profiles will shift. Developed nations, their growth slowed or stopped, will have more older citizens; new infections will attack primarily the immunosuppressed and subpopulations of the young, old, and poor. In developing countries, populations will be skewed strongly toward youth and early adulthood.

People's behavior and disease patterns are both influenced by whether they are young or old, prosperous or poor, cautious or impulsive. The young, who predominate in exploding populations, are far more given to risk-taking than the mature. Risky sex, drug use, and lack of education, along with mobility and rapid social change, have helped spread AIDS and other infections, and will continue to do so. And if the young lack food, jobs, and hope, they become social and medical time bombs, both victims and disseminators of disease.

In a best-case global scenario, population growth will slow because of two powerful contraceptives, mass education and a rising standard of living. Economic and agricultural development, like industrial development, will proceed in an atmosphere of social and political stability. Such improvement may seem a utopian hope, but it is already taking place in some developing nations. In some, unfortunately, these improvements are fragile, and many poor countries are falling further behind than ever. As geographer Robert Kates has said, we will survive in a tolerable environment if we "manage the transition to a warmer, more crowded, more connected but more diverse world."

The worst-case scenario goes beyond hunger, environmental degradation, and epidemics. It includes a massive global die-off caused by pandemics—perhaps by revived bubonic/pneumonic plague, a virulent new flu virus, a new airborne hemorrhagic fever, or germs that lurk undiscovered in other species. The Black Death killed one-quarter or more of humanity. It is conceivable that pandemics could do so again.

Some biologists argue that die-offs do not follow population explosions as winter does summer. In fact, they claim, epidemics rarely cause excess deaths in most species; these scientists make interesting statistical arguments to prove the case. It is true that human die-offs are not automatically triggered by population levels, as if by a thermostat. But for a variety of reasons, rising numbers often invite epidemics, as if to relieve intolerable pressure on both a species and its environment. In some species, they do so with grim regularity. Whether population control by epidemics lies ahead for our species is open to debate.

Such disaster is no certainty at any given time. Even a century ago, few people would have imagined that the earth could support its present population. With ingenuity or luck or both, we may stumble through another doubling of the population without correction by pandemics. It does not depend on us alone. If the study of emerging diseases shows anything, it is that hosts and parasites dance, however differently, to the same tune. They are constrained by each other and evolve together.

That brings us to the last item on the panel's list, microbial adaptation and change. Microbes reproduce and evolve quickly, adjusting to new hosts and new conditions. This is especially true of the flu virus, with its intricately variable surface proteins, and RNA viruses, which mutate with high frequency—as researchers have learned with frustration in trying to develop drugs and vaccines against AIDS.

We have only recently realized that microbes are not a limited array of permanently fixed species. They live in a blizzard of genetic material that alters their infectivity, virulence, and resistance. The strep A bacterium and the smallpox virus have historically changed in virulence, whether by mutation, genetic recombination, or gene transfer by viruses and plasmids. Newfound ability to produce damaging toxins has recently produced severe *E. coli* disease, toxic shock syndrome, and toxic-shock-like syndrome. A similar change has occurred in *Haemophilus influenzae.* This fly-borne bacterium, named when it was incorrectly thought to cause flu, actually causes meningitis and conjunctivitis, an inflammation of the membrane lining the eyelids. In 1984 in Brazil, a mutant strain attacked children and killed 70 percent of them. Thus arose Brazilian purpuric fever, which now may exist on several continents. A vaccine has become available, but it is far from universally used.

The history of human disease is in large part a history of such adaptations. Some of the most important recent changes are due to increased "microbial traffic," a product of alterations in population, technology, lifestyle, and travel. Other changes can be attributed to the defenses microbes have developed against antibiotics, thanks to their overuse in humans and livestock. Similar resistance to pesticides has arisen in many disease-carrying insects, partly because of misuse or overuse. The task of maintaining public health has also been made harder by the banning of many effective pesticides for real or only suspected environmental and health reasons.

Obviously the panel's six factors work in synergy, and in unpredictable ways. Once new technology introduces an unfamiliar microbe to humans, their travel may spread it. After war or disaster stops public health measures, sex behavior can help spread an emerging disease. After altered land use exposes people to new microbes, the people may carry them to other environments and pass them to new animal reservoirs there. We are almost never in control of such things, and we never have been. We react to problems, set off cascades of complex changes, and react to those in turn. Our foresight is rarely complete or reliable. Regardless of human attention and efforts, such biological processes continue. In 1930, Charles Nicolle, a French bacteriologist who won a Nobel Prize for his research on the epidemiology of typhus, wrote:

> Nature's attempts to create new diseases are as constant as they usually are vain. What happened in antiquity when, by exception, nature succeeded in an attempt is repeated at every moment now and will continue to be repeated always. It is inevitable. Equally inevitable is the fact that . . . when we become aware of these diseases, these are already fully formed.

Today the race between human ingenuity and that of microbes is tightening. The ingenuity of microbes lies in their genes, that of humans in their imagination and intelligence. The first step people must take is to recognize that change is a normal part of their relationship with microbes. Epidemiologist Robert Shope, a co-chairman of the 1992 panel, says, "The danger posed by infectious diseases has not gone away—it's worsening. Although we do not know where the next microbe or virus will appear . . . we know that new outbreaks are certain."

Some members of the panel are more optimistic than others. Karl Johnson, who did some of the original research on the Machupo and Reston viruses, says humans have already slashed and hacked their way into most of the world's unique ecosystems; few new fevers wait to emerge. Johnson may be right, but some of his colleagues are less confident, and Johnson agrees with the panel's suggestions. The appearance of new diseases is a process that sometimes accelerates, as in recent decades, and sometimes halts, but one must always assume it is happening or about to happen. The panel made these recommendations, among others:

- Set up a surge capacity for producing vaccines and drugs to combat epidemics, and facilities for storing and distributing them.

- Encourage development of new pesticides for use against disease vectors, with more input from epidemiologists.

- Improve disease-surveillance programs nationally and internationally. This includes standardizing the thousands of city, county, state, and federal systems for disease reporting in the United States.

- Create global surveillance systems for new diseases, emerging microbial drug resistance, and hospital infections.

- Train a new generation of experts in infectious diseases and epidemiology.

The panel's other co-chairman, microbiologist Joshua Lederberg, says that implementing all of the recommendations would cost less than half a billion dollars, and "even tens of millions would make a very big difference." It is not surprising that such action has not followed. While the emergence of one disease or another in the United States has caused alarm, the larger question of emerging diseases has not.

One reason is that some of the world's worst infectious killers seem distant. Malaria, schistosomiasis, sleeping sickness, and leprosy debilitate or kill millions of people each year, as do tuberculosis, influenza, and diarrheal diseases. Some of these are limited entirely or mainly to poor tropical and subtropical regions. What is not adequately understood is that in our time, because of mobility and high-speed travel and trade, no disease is irrelevant to the rest of the world.

Our society has been lulled also by sentimentality about nature. The new popular concern with environmental matters has indulged in an uncritical love for an "unspoiled" wilderness that in fact has not existed in much of the world for millennia. It does not include a realistic sense of parasitism as part of the web of life. In the altered nature in which we live, which we have helped create, species continue to evolve and make new relationships. Understanding ecology means not just sympathy for whales and owls but an appreciation of the entire biota, from humans to weeds to the smallest microscopic parasite.

Another reason for ignoring the present crisis of emerging diseases is to avoid the fear and pessimism it could engender. Actually, the situation calls for alertness, sometimes alarm, but not for panic or fatalism. History offers grounds for cautious optimism. Our relationship with the environment and with pathogens has always been one of sporadic crises. We are now in another major crisis, and history shows that such periods are coming closer and closer together.

This book began by asking where new diseases come from, and why so many are arriving now. The answer, as we have seen, is that their arrival is an ancient, natural process. It has risen to crisis proportions whenever people have made radical changes in their lifestyle and environment in order to live in greater plenty. Every age of new plenty demanded a price of biological readaptation. Diseases occurred in increased numbers when our ancestors left the trees for the ground; when nomads became hunters and spread around the world; when village life began, and with the growth of cities; with the start of global travel, and then with the Industrial Revolution; with the social and technological results of prosperity. The current wave of diseases has arisen with the acceleration of travel, technology, and social and environmental change in the past half-century. Attempts at plenty have a price, and always the human decision has been that the price is worth the struggle.

The twenty-first century will bring a savage test. Infectious diseases remain the world's leading cause of death; they will remain so for a long time to come. We will probably see more zoonoses, more mutated and drug-resistant germs, new microbe carriers, more environmental degradation and population pressures. Yet there is much in our favor. Neither we nor our pathogens would exist today without a huge capacity of mutual adaptation. We and they have survived cataclysmic change before. We have seen diseases arrive; we have seen them also become less deadly or disappear. The pace of change is faster than ever, but our tools for responding are better. We cure and control more diseases, and we are learning at a terrific pace about germs, genes, and immunity.

We know that we cannot flee disease by moving to a suburb or even a wilderness. There is no magic city encased in a protective bubble, nor will there ever be. Our most troubling vulnerability is our spotty record of serving our own best interests. Ignorance, greed, and shortsightedness often keep us from using the tools we have. Perhaps the prospect of global sickness will move us to protect ourselves better. Or perhaps we can rouse the same determination to deal with infectious disease that has informed recent efforts to curb pollution, environmental ruin, and harmful behavior. Then perhaps individuals and governments will be able to keep up with the wheel of change.

bibliography

This bibliography has two aims. One is to list the sources I found most useful and those quoted or mentioned in the text. The other is to guide readers to works they can consult if they want to read more, including many with extensive bibliographies and scholarly apparatus. Many of the sources are necessarily specialized, but when possible I have used references in easily available periodicals such as *Science*, *Scientific American*, *The New York Times*, and the *Morbidity and Mortality Weekly Report* (*MMWR*), published by the federal Centers for Disease Control.

A small core of works are sources for many parts of this book; they are listed first, in the Core Bibliography. I have sometimes pointed out which references are meant for specialists and which for general readers. Those recommended to "serious readers" do not require a specialist's knowledge, but they are easier for people with some scientific or scholarly background.

Finally, I have presumed to say that I consider some works indispensable, some particularly useful, and some brilliant or charming. It is the least they deserve.

CORE BIBLIOGRAPHY

Brothwell, Don, and A. T. Sandison, eds. (1967). *Diseases in Antiquity*. Springfield, IL: Thomas. Somewhat dated, but still indispensable to scholars.

Burnet, Macfarlane, and David White (1972). *Natural History of Infectious Disease* (4th ed). Cambridge: Cambridge Univ. Press. A classic, and readable.

Cartwright, Frederick (1974). *Disease and History*. New York: Signet. Lively. For general readers.

Cockburn, Aidan (1983). *The Evolution and Eradication of Infectious Diseases*. Westport, CT: Greenwood.

Croll, Neil, and John Cross, eds. (1983). *Human Ecology and Infectious Diseases*. New York: Academic Press.

Fiennes, Richard (1978). *Zoonoses and the Origins and Ecology of Human Disease*. New York: Academic Press.

Grmek, Mirko (1989). *Disease in the Ancient Greek World.* Baltimore: The Johns
  Hopkins Univ. Press. Thorough and thoughtful; covers far more than the title
  suggests. For scholars.
Hart, Gerald, ed. (1983). *Disease in Ancient Man.* Toronto: Clarke Irwin. For special-
  ists.
Kiple, Kenneth, ed. (1993). *The Cambridge World History of Human Disease.* New York:
  Cambridge Univ. Press. Useful for scholars.
Lederberg, Joshua, and Robert Shope, eds. (1992). *Emerging Infections: Microbial
  Threats to Health in the United States.* Washington, DC: National Academy Press.
  Indispensable. For scholars and health professionals.
Linton, Alan (1982). *Microbes, Man and Animals.* New York: Wiley. For specialists.
McEvedy, C., and R. Jones (1978). *Atlas of World Population History.* Harmondsworth,
  England: Penguin.
McKeown, Thomas (1988). *The Origins of Human Disease.* Oxford: Basil Blackwell. A
  demographic history. Excellent.
McNeill, William (1976). *Plagues and People.* Garden City, NY: Anchor. A classic. For
  specialists, serious readers, and anyone interested enough to look.
Morse, Stephen, ed. (1993). *Emerging Viruses.* New York: Oxford Univ. Press. Impor-
  tant for specialists.
Ranger, Terence, and Paul Slack, eds. (1992). *Epidemics and Ideas: Essays on the
  Historical Perception of Pestilence.* Cambridge: Cambridge Univ. Press.
Russell, W. M. S. (1983). "The Palaeodemographic View." In Hart (see above),
  pp. 217–253.
Stanley, N. F., and R. A. Joske, eds. (1980). *Changing Disease Patterns and Human
  Behaviour.* New York: Academic Press. Indispensable collection of papers for
  specialists and serious readers.
White, David, and Frank Fenner (1986). *Medical Virology* (3rd ed.). New York:
  Academic Press.
Wood, Corinne (1979). *Human Sickness and Health: A Biocultural View.* Mountain
  View, CA: Mayfield. Intended as a college text, but a good introduction to
  medical anthropology for serious nonspecialists.

CHAPTER ONE

Almost every incident and disease mentioned in this chapter is discussed at greater
length in a later chapter, for which references are provided. Several works in the Core
Bibliography address the central topics, especially those by Burnet and White,
Fiennes, Lederberg and Shope, Morse, and Stanley and Joske.

Brandt, Allan (1987). *No Magic Bullet.* New York: Oxford Univ. Press. A good social
  history of STDs in the United States.
Crosby, Alfred (1989). *America's Forgotten Pandemic: The Influenza of 1918.* Cambridge:
  Cambridge Univ. Press. Well written.
Dubos, René (1965). *Man Adapting.* New Haven, CT: Yale Univ. Press. Though dated
  in places, still a good introduction to the evolving relationship of humans,
  microbes, and the environment.
Garrett, Laurie (1994). *The Coming Plague.* New York. Farrar, Straus & Giroux.
  Useful, but long and disorganized.

Gibbons, Ann (1993). "Where Are 'New' Diseases Born?" *Science*, 261, pp. 680–681.

"Global Change" (1992). *Science*, 256, pp. 1138–1147. A good group of articles on planetary warming and its biological effects.

Henig, Robin (1993). *A Dancing Matrix: Voyages Along the Viral Frontier.* New York: Knopf. Good introduction for general readers.

Hoyle, Fred, and M. C. Wickramasinghe (1979). *Diseases from Space.* London: Dent.

"Infectious Diseases on the Rebound in the U.S., a Report Says." *The New York Times,* May 10, 1994.

"Lesson of the Plague. Beware of 'Vanquished' Diseases." *The New York Times,* Sept. 27, 1994.

Levins, Richard, et al. (1994). *American Scientist*, 82(2), pp. 52–60.

Mack, Arien, ed. (1991). *In Time of Plague: The History and Social Consequences of Lethal Epidemic Disease.* New York: New York Univ. Press. Contains some interesting essays.

Mitchison, Avrion (1993). "Will We Survive?" *Scientific American*, 269(3), pp. 136–144.

Radetsky, Peter (1991). *The Invisible Invaders.* Boston: Little, Brown.

Surgeon General (1979). *Healthy People: The Surgeon General's Report on Health Promotion and Disease Prevention.* Washington, DC: U.S. Department of Health, Education and Welfare.

"Threat Perceived from Emerging Microbes" (1992). *Science News*, 142(17), p. 278.

CHAPTER TWO

I can touch only briefly here on the rich and quickly changing literatures on the origins of life, symbiosis, and hominid evolution. There is good general background in the Core Bibliography works of Burnet and White, Cockburn, Fiennes, Hart, and Wood.

Blumenschine, Robert, and John Cavallo (1992). "Scavenging and Human Evolution." *Scientific American*, 267(4), pp. 90–96.

Chyba, Christopher, et al. (1990). "Cometary Delivery of Organic Molecules to the Early Earth." *Science*, 249, pp. 366–373.

Crick, F. H. C. (1981). *Life Itself: Its Origin and Nature.* New York: Simon & Schuster.

Ebert, Dieter (1994). "Virulence and Local Adaptation of a Horizontally Transmitted Parasite." *Science*, 265, pp. 1084–1085.

Edison, Millicent, et al. (1988). "Feline Plague in New Mexico: Risk Factors and Transmission to Humans." *American Journal of Public Health*, 78(10), pp. 1333–1335.

Fenner, Frank (1980). "Sociocultural Change and Environmental Disease." In Stanley and Joske (see Core Bibliography), pp. 7–26.

Fiennes, Richard (1967). *Zoonoses of the Primates.* Ithaca, NY: Cornell Univ. Press.

Golley, Frank (1994). *A History of the Ecosystem Concept in Ecology.* New Haven, CT: Yale Univ. Press.

Gregg, Charles (1985). *Plague: An Ancient Disease in the Twentieth Century.* Albuquerque: Univ. of New Mexico Press. Readable.

Hamilton, William D. (1988). "Sex and Disease." In Robert Bellig and George Stevens, eds., *Nobel Conferences XXIII: The Evolution of Sex.* New York: Harper &

Row, pp. 65–95. Presents the interesting theory that sexual reproduction evolved partly in response to infectious disease.

Hare, Ronald (1967). "The Antiquity of Diseases Caused by Bacteria and Viruses." In Brothwell and Sandison (see Core Bibliography), pp. 115–131. A useful essay.

Jonas, David, and Doris Klein (1970). *Man-Child.* New York: McGraw-Hill. On regressive evolution.

Khakhina, Liya N. (1993). *Concepts of Symbiogenesis* (tr. Stephanie Merkel and Robert Coalson). New Haven, CT: Yale Univ. Press. On early Russian symbiosis theorists. For scholars.

Leakey, M. D., and J. M. Harris, eds. (1987). *Laetoli: A Pliocine Site in Northern Tanzania.* New York: Oxford Univ. Press.

Lipps, Jere (1993). *Fossil Prokaryotes and Protists.* Cambridge, MA: Blackwell Scientific.

Maino, Guido (1975). *The Healing Hand: Man and Wound in the Ancient World.* Cambridge, MA: Harvard Univ. Press.

Margulis, Lynn, and René Fester, eds. (1991). *Symbiosis as a Source of Evolutionary Innovation.* Cambridge, MA: MIT Press.

Margulis, Lynn, and D. Sagan (1986). *Microcosmos: Four Billion Years of Evolution from Our Microbial Ancestors.* New York: Summit.

Morse, Stephen, ed. (1993). *The Evolutionary Biology of Viruses.* New York: Raven.

Ortner, D. J., and W. G. Putschar (1981). *Identification of Pathological Conditions in Human Skeletal Remains (Smithsonian Contributions to Anthropology,* 28). Washington, DC: Smithsonian Institution Press.

Rebek, Julius, Jr. (1994). "Synthetic Self-Replicating Molecules." *Scientific American,* 271(1), pp. 48–55.

Rennie, John (1992). "Trends in Parasitology: Living Together." *Scientific American,* 266(1), pp. 122–133.

Solecki, Ralph (1971). *Shanidar: The First Flower People.* New York: Knopf. For general readers.

Stehr-Green, Jeanette, and Peter Schantz (1986). "Trichinosis in Southeast Asian Refugees in the United States." *American Journal of Public Health* 76(10), pp. 1238–1239.

Stringer, C. B., and P. Andrews (1988). "Genetic and Fossil Evidence for the Origins of Modern Humans." *Science,* 239, pp. 1263–1268. The African Eve theory.

Swinton, W. E. (1983). "Animal Paleopathology: Its Relation to Ancient Human Disease." In Hart (see Core Bibliography), pp. 50–60.

Toft, Catherine, André Aeschlimann, and Liana Bolis, eds. (1991). *Parasite-Host Associations: Coexistence or Conflict?* Oxford: Oxford Univ. Press.

Trinkaus, Erik, and Pat Shipman (1994). *The Neandertals.* New York: Vintage. For serious readers. A fascinating description of how intellectual and social fashions have shaped interpretations of early humans.

Yoon, Carol K. (1993). "What Might Cause Parasites to Become More Virulent?" *Science,* 259, p. 1402.

Young, D. (1973). "Was There an Unsuspected Killer Aboard 'The Unicorn'?" *The Beaver,* 304(3), pp. 9–15.

Zimmerman, Michael (1981). "Homo erectus and Hypervitaminosis A." *Papers Presented at the Annual Meeting of the Paleopathology Association 22 April 1981,* p. D2.

CHAPTER THREE

Of the books in the Core Bibliography, Brothwell and Sandison, Burnet and White, Croll and Cross, Grmek, McNeill, and Stanley and Joske contain interesting details on subjects discussed in this chapter. For the scholar or ambitious nonspecialist, Cohen's book (see below) is a mine of information, with a good bibliography. The *Paleopathology Newsletter* and other publications of the Paleopathology Association are indispensable to scholars. Unfortunately, general readers find little in this area that is thorough and readable outside such periodicals as *Scientific American*. Tell Abu Hureyra is covered by Legge and Rowley-Conway, Molleson, and Moore.

Angel, J. L. (1966). "Porotic Hyperostosis, Anemias, Malarias, and Marshes in the Prehistoric Eastern Mediterranean." *Science,* 153, pp. 760–763.

Benfer, Robert (1984). "The Challenges and Rewards of Sedentism: The Preceramic Villages of Paloma, Peru." In Cohen and Armelagos (see below), pp. 531–555.

Black, F. L. (1980). "Modern Isolated Pre-Agricultural Populations as a Source of Information on Prehistoric Epidemic Patterns." In Stanley and Joske (see Core Bibliography), pp. 37–54.

Cann, R. L., M. Stoneking, and A. C. Wilson (1987). "Mitochondrial DNA and Human Evolution." *Nature,* 325, pp. 31–36. The original African Eve proposition.

Cavalli-Sforza, L., P. Menozzi, and A. Piazza (1993). "Demic Expansions and Human Evolution." *Science,* 259, pp. 639–646.

Cohen, Mark N. (1989). *Health and the Rise of Civilization.* New Haven, CT: Yale Univ. Press. Highly recommended for scholars and serious readers.

———, and G. J. Armelagos (1984). *Paleopathology at the Origins of Agriculture.* New York: Academic Press.

Cross, John H., and Manoon Bhaibulaya (1983). "Intestinal Capillariasis in the Philippines and Thailand." In Croll and Cross (see Core Bibliography), pp. 103–136. The story of the Mystery Disease of Pudoc.

El-Najjar, M. Y. (1976). "Maize, Malaria, and the Anemias in the Pre-Columbian New World." *Yearbook of Physical Anthropology,* 20, pp. 329–337.

Elphinstone, J. J. (1971). "The Health of Aborigines with No Previous Association with Europeans." *Medical Journal of Australia,* 2, pp. 293–303.

Gilbert, R. I., and J. H. Mielke (1985). *The Analysis of Prehistoric Diets.* New York: Academic Press.

Haynes, Gary (1991). *Mammoths, Mastodonts, and Elephants.* Cambridge: Cambridge Univ. Press. Discusses the idea of extinction by hunting.

Kliks, Michael (1983). "Paleoparasitology." In Croll and Cross (see Core Bibliography), pp. 291–313.

——— (1990). "Helminths as Heirlooms and Souvenirs: A Review of New World Parasitology." *Parasitology Today,* 6(4), pp. 93–100.

Lee, R. B., and I. DeVore, eds. (1968). *Man the Hunter.* Chicago: Aldine.

——— (1976). *Kalahari Hunter-Gatherers.* Cambridge, MA: Harvard Univ. Press.

Legge, Anthony, and Peter Rowley-Conway (1987). "Gazelle Killing in Stone Age Syria." *Scientific American,* 257, pp. 88–95.

Lewin, Roger (1986). "Polynesians' Litter Gives Clues to Islands' History." *Science,* 231, pp. 453–454.

Lorenz, Konrad (1964). *Man Meets Dog* (tr. Marjorie Wilson). Harmondsworth, England: Penguin. A charming book by a brilliant ethologist. For any reader.

Mims, Cedric (1980). "The Emergence of New Infectious Diseases." In Stanley and Joske (see Core Bibliography), pp. 231–250.

Molleson, Theya (1994). "The Eloquent Bones of Abu Hureyra." *Scientific American,* 271(2), pp. 70–75. On clues to occupational illnesses.

Moore, Andrew (1979). "A Pre-Neolithic Farmers' Village on the Euphrates." *Scientific American,* 241(2), pp. 62–70.

Morrell, Virginia (1990). "Confusion in Earliest America." *Science,* 17, pp. 439–441.

Myers, F. R. (1988). "Critical Trends in the Study of Hunter-Gatherers." *Annual Review of Anthropology,* 17, pp. 261–282.

Palkovich, A. M. (1987). "Endemic Disease Patterns in Paleopathology: Porotic Hyperostosis." *American Journal of Physical Anthropology,* 74(4), pp. 527–537.

Reinhard, K. J. (1991). "Recent Contributions to New World Archaeoparasitology." *Parasitology Today,* 7(4), pp. 81–82.

Rose, J. C., B. Burnett, and A. M. Harmon (1990). "Infections and Ecology in the Trans-Mississippi South, USA." *Papers on Paleopathology Presented at the Eighth European Members Meeting, Paleopathology Association,* p. 19.

Simmons, Alan H., et al. (1988). " 'Ain Ghazal: A Major Neolithic Settlement in Central Jordan." *Science,* 240, pp. 35–39.

Smith, Bruce D. (1992). *Eastern Origins: Essays on the Origins of Agriculture in Eastern North America.* Washington, DC: Smithsonian Institution Press.

Steinbock, R. Ted (1976). *Paleopathological Diagnosis and Interpretation.* Springfield, IL: Thomas.

Straus, Lawrence (1985). "Stone Age Prehistory of Northern Spain." *Science,* 230, pp. 501–507.

Stringer, C. B., and P. Andrews (1988). "Genetic Fossil Evidence for the Origins of Modern Humans." *Science,* 239, pp. 1263–1268. Supporting the African Eve theory.

White, Randall (1986). *Dark Caves, Bright Visions: Life in Ice Age Europe.* New York: Norton.

Wolpoff, Milford, and Alan Thorne (1991). "The Case Against Eve." *New Scientist,* 130(1774), pp. 37–41.

CHAPTER FOUR

This chapter draws on the same core sources as does chapter 3, and also on Burnet and White, Cartwright, Cockburn, Fiennes, Hart, Kiple, Linton, McKeown, Russell, and Wood. The general reader can get good overviews from Cartwright, Crosby (see below), and McNeill.

Ancient Cities (1994). *Scientific American* (Special Issue). Very useful to nonspecialists.

Angel, J. L. (1971). *The People of Lerna.* Washington, DC: Smithsonian Institution Press.

Bollet, Alfred (1987). *Plagues and Poxes: The Rise and Fall of Epidemic Disease.* New York: Demos. A collection of readable essays.

Bruce-Chwatt, Leonard, and Julian de Zulueta (1980). *The Rise and Fall of Malaria in Europe*. New York: Oxford Univ. Press. Recommended for scholars.

Cockburn, Aidan, and Eve Cockburn, eds. (1980). *Mummies, Disease, and Ancient Cultures*. Cambridge: Cambridge Univ. Press. Fascinating. Explains the techniques of paleopathology.

Cohen, Mark N. (1989). See above, chap. 3.

Crosby, Alfred (1986). *Ecological Imperialism: The Biological Expansion of Europe, 900–1900*. New York: Cambridge Univ. Press. See especially chap. 2. Excellent book.

Fenner, Frank (1980). See above, chap. 2.

Goldman, L., et al. (1983). "The Value of the Autopsy in Three Medical Eras." *The New England Journal of Medicine*, 308, pp. 1000–1005.

Hare, Ronald (1967). See above, chap. 2.

Hopkins, D. R. (1983). *Princes and Peasants: Smallpox in History*. Chicago: Univ. of Chicago Press. Readable. For both scholars and nonspecialists.

Karlen, Arno (1984). *Napoleon's Glands and Other Ventures in Biohistory*. Boston: Little, Brown. See chaps. 4 and 5.

Langmuir, A. D., and C. G. Ray (1985). "The Thucydides Syndrome: A New Hypothesis for the Cause of the Plague of Athens." *The New England Journal of Medicine*, 313, pp. 1027–1030.

Longrigg, James (1980). "The Great Plague of Athens." *History of Science*, 18, pp. 209–225.

Mascie-Taylor, C. G. N., ed. (1993). *The Anthropology of Disease*. Oxford: Oxford Univ. Press.

Ortner, Donald, and Gretchen Theobald (1993). "Diseases in the Pre-Roman World." In Kiple (see Core Bibliography), pp. 247–261.

Pritchard, James B., ed. (1969). *Ancient Near Eastern Texts Relating to the Old Testament*. Princeton, NJ: Princeton Univ. Press.

Quétel, Claude (1990). *History of Syphilis* (tr. Judith Braddock and Brian Pike). Baltimore: The Johns Hopkins Univ. Press.

Rothberg, R. I., and T. K. Rabb (1983). *Hunger and History*. Cambridge: Cambridge Univ. Press.

"Skeletons Record the Burdens of Work." *The New York Times*, Oct. 27, 1987.

Stannard, Jerry (1993). "Diseases of Western Antiquity." In Kiple (see Core Bibliography), pp. 262–269.

Sussman, Max (1967). "Diseases in the Bible and the Talmud." In Brothwell and Sandison (see Core Bibliography), pp. 209–221.

Thucydides (1988). *History of the Peloponnesian War, Book 2* (tr. P. J. Rhodes). Warminster, England: Aris & Phillips.

CHAPTER FIVE

In the Core Bibliography, see Brothwell and Sandison, Cartwright, Cockburn, Fiennes, Grmek, Kiple, McKeown, McEvedy and Jones, McNeill, and Russell. Kiple and McNeill are especially useful for areas beyond Europe.

Anderson, R. M., and R. M. May, eds. (1982). *Population Biology of Infectious Diseases*. New York: Springer.

Biraben, J.-N. (1975). *Les hommes et la peste.* The Hague: Mouton. A major scholarly work.

Bruce-Chwatt, Leonard, and Julian de Zulueta (1980). See above, chap. 4.

"Children's Cemetery a Clue to Malaria as Rome Declined." *The New York Times,* July 26, 1994.

Gibbon, Edward (1957–1962). *The Decline and Fall of the Roman Empire.* New York: Dutton.

Gwei-Djen, Lu, and Joseph Needham (1993). "Diseases of Antiquity in China." In Kiple (see Core Bibliography), pp. 345–354.

Hopkins, D. R. (1983). See above, chap. 4.

Malthus, Thomas (1960). *On Population* (ed. Gertrude Himmelfarb). New York: Modern Library.

Moore, Patrick, and Claire Broome (1994). "Cerebrospinal Meningitis Epidemics." *Scientific American,* 271(5), pp. 38–45. Further understanding of how climate, environment, and human behavior interact to cause pandemics.

O'Neill, Ynez (1993). "Diseases of the Middle Ages." In Kiple (see Core Bibliography), pp. 270–278.

Patrick, Adam (1967). "Disease in Antiquity: Ancient Greece and Rome." In Brothwell and Sandison (see Core Bibliography), pp. 238–246.

Procopius (1914). *History of the Wars, I* (ed. H. B. Dewing). New York: Macmillan.

Selye, Hans (1956). *The Stress of Life.* New York: McGraw-Hill.

Turner, Michael, ed. (1986). *Malthus and His Time.* New York: St. Martin's.

Welch, William J. (1993). "How Cells Respond to Stress." *Scientific American,* 268(5), pp. 56–64.

Ziegler, Philip (1970). *The Black Death.* Harmondsworth, England: Penguin. A good introduction, and a useful bibliography.

CHAPTER SIX

Core references are Brothwell and Sandison, Cartwright, Grmek, Hart, Kiple, McEvedy and Jones, McKeown, McNeill, Ranger and Slack, Russell, and Wood. Gottfried is best for general readers on the Black Death, Biraben for scholars. Defoe offers perhaps the best picture of daily life during pestilence. Agnolo di Tura is quoted in Bowsky, Plutarch in Grmek.

Amelang, James, ed. and tr. (1991). *A Journal of the Plague Year: The Diary of the Barcelona Tanner Miquel Parets, 1651.* New York: Oxford Univ. Press.

Biraben, J.-N. (1975). See above, chap. 5.

Bowsky, William, ed. (1971). *The Black Death.* New York: Rinehart & Winston.

Creighton, Charles (1965). *History of Epidemics in Britain A.D. 664–1666* (2nd ed.). New York: Barnes & Noble. A classic of Victorian scholarship; still important.

Deaux, George (1969). *The Black Death, 1347.* New York: Weybright and Talley. For general readers.

Defoe, Daniel (1960). *A Journal of the Plague Year.* New York: Signet. A must, and a pleasure.

Dols, Michael (1977). *The Black Death in the Middle East.* Princeton, NJ: Princeton Univ. Press.

Dubos, René, and Jean Dubos (1953). *The White Plague: Tuberculosis, Man, and Society.* New Brunswick, NJ: Rutgers Univ. Press.

Fenner, Frank (1980). See above, chap. 2.

Gottfried, Robert (1983). *The Black Death.* New York: Free Press. First-rate, with a good bibliography; good introduction for a nonspecialist.

Gregg, Charles (1985). See above, chap. 2.

Hendrickson, Robert (1983). *More Cunning Than Man: A Social History of Rats and Men.* New York: Dorset. Interesting and readable.

Møller-Christenson, Vilhelm (1983). "Leprosy and Tuberculosis." In Hart (see Core Bibliography), pp. 129–138.

Morse, Dan (1967). "Tuberculosis." In Brothwell and Sandison (see Core Bibliography), pp. 249–271.

Robinson, Victor (1943). *The Story of Medicine.* New York: New Home Library.

Schultz, Michael (1981). "Aseptic Bone Necrosis Found in the Skeletal Material from the Merovingian Cemetery of Kleinlangheim, Southern Germany." *Paleopathology Newsletter,* no. 34, pp. 7–8.

Slack, Paul (1983). "The Social Effects of Plague in Early Modern Europe." In Hart (see Core Bibliography), pp. 254–262.

Steinbock, R. Ted (1976). See above, chap. 3.

Ziegler, Philip (1970). See above, chap. 5.

Zinsser, Hans (1935). *Rats, Lice and History.* Boston: Little, Brown. Though now dated in some places, this book is still informative and charming.

CHAPTER SEVEN

Core Bibliography sources for this chapter are Cartwright, Hart, Kiple, Linton, McEvedy and Jones, McNeill, Ranger and Slack, Russell, White and Fenner, and Wood. Crosby's two books (cited below) are excellent for general as well as scholarly readers. I have dipped only shallowly into the vast and fascinating literatures on New World exploration and pre-Columbian cultures.

Black, F. L. (1982). "Why Did They Die?" *Science,* 258, pp. 1739–1740.

Buikstra, Jane (1993). "Diseases of the Pre-Columbian Americans." In Kiple (see Core Bibliography), pp. 305–317.

"Chagas' Disease Claimed an Eminent Victim" (letter). *The New York Times,* June 15, 1989. On Darwin.

Cohen, I. Bernard (1992). "What Columbus 'Saw' in 1492." *Scientific American,* 256(6), pp. 100–106. Good essay.

Cohen, Mark N. (1989). See above, chap. 3.

———, and G. J. Armelagos (1984). See above, chap. 3.

Crosby, Alfred. (1972). *The Columbian Exchange.* Westport, CT: Greenwood. A learned and well-written book.

——— (1986). See above, chap. 4.

Dobyns, Henry (1963). "An Outline of Andean Epidemic History to 1720." *Bulletin of the History of Medicine,* 37, pp. 493–515.

——— (1983). *Their Numbers Become Thinned: Native American Population Dynamics in Eastern North America.* Knoxville: Univ. of Tennessee Press.

Duffy, John (1953). *Epidemics in Colonial America.* Baton Rouge: Louisiana State Univ. Press.

Goodman, A. H., and G. J. Armelagos (1985). "Disease and Death at Dr. Dickson's Mounds." *Natural History,* 94, pp. 12–18.

Hackett, C. J. (1978). "Treponematosis (Yaws and Treponarid) in Exhumed Australian Aboriginal Bones." *Records of the South Australia Museum,* 17, pp. 387–405.

Hopkins, D. R. (1983). See above, chap. 4.

Horgan, John (1992). "Early Arrivals: Scientists Argue over How Old the New World Is." *Scientific American,* 266(2), pp. 17, 20.

"In Thatched Roofs of Argentine Poor, an Insect That Saps Health of Millions." *The New York Times,* June 4, 1990. On Chagas' disease.

Kilbourne, Edwin (1987). *Influenza.* New York: Plenum. An excellent reference work by one of the world's leading experts on flu.

Kiple, Kenneth (1984). *The Carribean Slave Trade: A Biological History.* Cambridge, England: Cambridge Univ. Press.

Kliks, Michael (1990). See above, chap. 3.

Merbs, Charles F. (1983). "Paleopathology in North America: A Regional Survey." *Papers on Paleopathology Presented at the Tenth Annual Meeting, Paleopathology Association,* pp. 1–6.

———— (1992). "A New World of Infectious Diseases." *Yearbook of Physical Anthropology,* 35, pp. 3–42.

Patterson, K. David (1986). *Pandemic Influenza, 1700–1900.* Totowa, NJ: Rowman & Littlefield.

Quétel, Claude (1990). See above, chap. 4.

Ramenofsky, Ann (1993). "Diseases of the Americas, 1492–1700." In Kiple (see Core Bibliography), pp. 317–328.

Reinhard, K. J. (1991). See above chap. 3.

Rothammer, F., et al. (1985). "Chagas' Disease in Pre-Columbian South America." *American Journal of Physical Anthropology,* 68(4), pp. 495–507.

Rouse, Irving (1992). *The Tainos.* New Haven, CT: Yale Univ. Press.

Sulzer, A. J., et al. (1978). "Study of Coinciding Foci of Malaria and Leptospirosis in the Peruvian Amazon Area." *Transactions of the Royal Society of Tropical Medicine & Hygiene,* 72, pp. 76–83.

Verano, John, and Douglas Ubelaker, eds. (1992). *Disease and Demography in the Americas.* Washington, DC: Smithsonian Institution Press.

Wood, W. Barry (1961). *From Miasmas to Molecules.* New York: Columbia Univ. Press.

Zulueta, Julian de (1980). "Man and Malaria." In Stanley and Joske (see Core Bibliography), pp. 175–186.

CHAPTER EIGHT

In the Core Bibliography see especially Croll and Cross, Fiennes, Grmek, Linton, McNeill, Ranger and Slack, and Wood.

Baker, B. J., and G. J. Armelagos (1968). "The Origin and Antiquity of Syphilis." *Current Anthropology,* 29(5), pp. 703–737.

Brandt, Allan (1988). "The Syphilis Epidemic and Its Relation to AIDS." *Science,* 239, pp. 375–380.

Busvine, James (1976). *Disease Transmission by Insects.* New York: Springer. Authoritative.

———— (1976). *Insects, Hygiene and History.* London: Athlone.

Cloudsley-Thompson, J. L. (1976). *Insects and History.* New York: St. Martin's.

Edwards, R. Dudley, and T. Desmond Williams, eds. (1957). *The Great Famine.* New York: New York Univ. Press.

El-Najjar, M. Y. (1979). "Human Treponematosis and Tuberculosis: Evidence from the New World." *American Journal of Physical Anthropology,* 51, pp. 599–618.

Farhang-Azad, A., R. Traud, and S. Baqar (1985). "Transovarial Transmission of Murine Typhus Rickettsiae in *Xenopsylla cheopis* Fleas." *Science,* 227, pp. 543–545.

Fracastoro, Girolamo (1930). *Contagion* (tr. W. C. Wright). New York: Putnam.

———— (1934). *The Sinister Shepherd* (tr. William Van Wyck). Los Angeles: Primavera.

Goff, C. W. (1967). "Syphilis." In Brothwell and Sandison (see Core Bibliography), pp. 279–294.

Hackett, C. J. (1967). "The Human Treponematoses." In Brothwell and Sandison (see Core Bibliography), pp. 152–169.

———— (1983). "Problems in the Palaeopathology of the Human Treponematoses." In Hart (see Core Bibliography), pp. 106–128.

Hendrickson, Robert (1983). See above, chap. 6.

Hudson, E. H. (1965). "Treponematosis and Man's Social Evolution." *American Anthropologist,* 67, pp. 885–901.

———— (1968). "Christopher Columbus and the History of Syphilis." *Acta Tropica, 25,* pp. 1–16.

Ladurie, Emmanuel Le Roy (1988). *Times of Feast, Times of Famine: A History of Climate Since the Year 1000* (tr. Barbara Bray). New York: Farrar, Straus & Giroux.

Marsden, P. D. (1983). "The Transmission of *Trypanosoma cruzi* Infection to Man and Its Control." In Croll and Cross (see Core Bibliography), pp. 253–289.

Moffat, Anne Simon (1992). "Improving Plant Disease Resistance." *Science,* 257, pp. 482–483.

Ó Gráda, Cormac (1989). *The Great Irish Famine.* London: Macmillan. Scholarly and judicious.

Post, John (1985). *Food Shortage, Climatic Variability, and Epidemic Disease in Preindustrial Europe: The Mortality Peak in the Early 1740s.* Ithaca, NY: Cornell Univ. Press.

Quétel, Claude (1990). See above, chap. 4.

Rosebury, Theodor (1973). *Microbes and Morals.* New York: Ballantine. A well-written social and biological history of STDs, from the pre-AIDS era.

Rotberg, Robert, and Theodore Rabb, eds. (1985). *Hunger and History.* Cambridge: Cambridge Univ. Press.

Sandison, A. T., and Calvin Wells (1967). "Diseases of the Reproductive System." In Brothwell and Sandison (see Core Bibliography), pp. 498–520.

"A Virulent Potato Fungus Is Killing the Northeast Crop." *The New York Times,* Nov. 12, 1994.

Woodham-Smith, Cecil (1962). *The Great Hunger.* New York: Harper & Row. A detailed and vivid account.

Zinsser, Hans (1935). See above, chap. 6. The classic history of typhus and related diseases.

CHAPTER NINE

In the Core Bibliography see Cartwright, Cockburn, Kiple, McEvedy and Jones, McKeown, Russell, Stanley and Joske, and especially Ranger and Slack. These and many works listed below have bibliographies on the interesting histories of cholera, public health, and epidemiology.

Barua, Dhiman, and William Greenough, eds. (1992). *Cholera.* New York: Plenum Medical.

Bollet, Alfred (1987). See above, chap. 4. Contains the Balto story.

Crosby, Alfred (1989). See above, chap. 1.

De Kruif, Paul (1926). *Microbe Hunters.* New York: Harcourt, Brace. This best-seller of the 1920s still conveys powerfully the excitement and optimism of the triumphs of pioneering bacteriology.

Duffy, John (1990). *The Sanitarians: A History of American Public Health.* Urbana, IL: Univ. of Illinois Press.

Durey, Michael (1979). *The Return of the Plague: British Society and the Cholera 1831–2.* New York: Humanities Press.

Evans, Richard J. (1992). "Epidemics and Revolutions: Cholera in Nineteenth-Century Europe." In Ranger and Slack (see Core Bibliography), pp. 149–174.

Hare, Ronald (1967). See above, chap. 2.

Hoehling, A. A. (1961). *The Great Epidemic.* Boston: Little, Brown. On influenza.

Hopkins, D. R. (1983). See above, chap. 4.

Howard-Jones, Norman (1980). "Prelude to Modern Preventive Medicine." In Stanley and Joske (see Core Bibliography), pp. 69–80.

Joske, R. A. (1980). "The Physician and Changing Patterns of Human Disease and Death." In Stanley and Joske (see Core Bibliography), pp. 551–566.

Kilbourne, Edwin (1987). See above, chap. 7.

Mack, Arien (1991). See above, chap. 1.

Patterson, K. David (1986). See above, chap. 7.

Pickstone, John (1992). "Dearth, Dirt and Fever Epidemics: Rewriting the History of British 'Public Health,' 1780–1850." In Ranger and Slack (see Core Bibliography), pp. 125–148.

Porter, Katherine Anne (1939). *Pale Horse, Pale Rider.* New York: Harcourt, Brace. Splendid.

Rosenberg, Charles (1987). *The Cholera Years: The United States in 1832, 1849, and 1866.* Chicago: Univ. of Chicago Press. An excellent book for scholars and general readers.

Wohl, Anthony (1983). *Endangered Lives: Public Health in Victorian Britain.* London: Dent.

Wood, W. Barry (1961). See above, chap. 7.

CHAPTER TEN

In the Core Bibliography, see Fiennes, Kiple, Lederberg and Shope, Linton, White and Fenner, and especially Morse.

Barinaga, Marcia (1993). "Satellite Data Rocket Disease Control Efforts into Orbit." *Science,* 261, pp. 31–32.

Benison, Saul (1972). "The History of Polio Research in the United States." In Gerald Holton, ed., *The Twentieth Century Sciences.* New York: Norton, pp. 308–343.

Busvine, James (1976). See above, chap. 8.

Centers for Disease Control (1988). "Arboviral Infections of the Central Nervous System—United States, 1987." *MMWR,* 37(33), pp. 506–515.

———— (1988). "Management of Patients with Suspected Viral Hemorrhagic Fever." *MMWR Supplement,* 37(S-3), pp. 1–15.

———— (1989). "Dengue Epidemic—Ecuador, 1988." *MMWR,* 38(24), pp. 419–421.

———— (1993). "Hantavirus Infection—Southwestern United States: Interim Recommendations for Risk Reduction." *MMWR,* 42(RR-11).

———— (1990). "Rocky Mountain Spotted Fever and Human Ehrlichiosis—United States, 1989." *MMWR,* 39(17), pp. 281–284.

———— (1990). "Update: Ebola-Related Filovirus Infection in Nonhuman Primates." *MMWR,* 39(2), pp. 22–30.

———— (1990). "Yellow Fever Vaccine." *MMWR: Recommendations and Reports,* 39(RR-6), pp. 1–6.

Cooper, J. I., and F. O. MacCallum (1984). *Viruses and the Environment.* London: Chapman & Hall.

"Encephalitis Reports at End, Florida Says Outbreak Is Over." *The New York Times,* December 21, 1990.

Fenner, Frank, et al. (1988). *Smallpox and Its Eradication.* Geneva: World Health Organization.

Francy, D. B., et al. (1990). "A New Arbovirus from *Aedes albopictus.*" *Science,* 250, pp. 1738–1740.

Fuller, John (1974). *Fever! The Hunt for a New Killer Virus.* New York: Reader's Digest Press. Lively account of Lassa fever.

Gibbons, Ann (1991). "Saying So Long to Polio." *Science,* 251, p. 1020.

Halstead, Scott (1988). "Pathogenesis of Dengue." *Science,* 239, pp. 476–481.

Hughs, James, et al. (1993). "Hantavirus Pulmonary Syndrome: An Emerging Infectious Disease." *Science,* 262, pp. 850–851.

Johnson, Karl (1993). "Emerging Viruses in Context: An Overview of Viral Hemorrhagic Fevers." In Morse (see Core Bibliography), pp. 46–57. Good overview.

"Killer Disease, Borne by Rodents, Is Found in Wider Areas of U.S." *The Wall Street Journal,* January 14, 1994.

Kraut, Alan M. (1994). *Silent Travelers: Germs, Genes, and the "Immigrant Menace."* New York: Basic Books.

Le Duc, James, et al. (1993). "Hantaan (Korean Hemorrhagic Fever) and Related Rodent Zoonoses." In Morse (see Core Bibliography), pp. 149–158.

Linthicum, Kenneth, et al. (1987). "Detection of Rift Valley Fever Viral Activity in Kenya by Satellite Remote Sensing Imagery." *Science,* 235, pp. 1656–1659.

Marshall, Eliot (1993). "Hantavirus Outbreak Yields to PCR." *Science,* 262, pp. 832–836.

Mitchell, C. J., et al. (1992). "Isolation of Eastern Equine Encephalitis Virus from *Aedes albopictus* in Florida." *Science,* 257, pp. 526–527.

Monath, Thomas (1993). "Arthropod-Borne Viruses." In Morse (see Core Bibliography), pp. 138–148.

"Mosquito Is Linked to Deadly Virus." *The New York Times,* July 24, 1992.

Nichol, Stuart, et al. (1993). "Genetic Identification of a Hantavirus Associated with an Outbreak of Acute Respiratory Illness." *Science,* 262, pp. 914–917.

"Outbreak of Polio Alarms Officials." *The New York Times,* October 8, 1991.

Pinheiro, Francisco, et al. (1982). "Transmission of Oropouche Virus from Man to Hamster by the Midge *Culicoides paraensis.*" *Science,* 215, pp. 1251–1252.

"Polio Synthesized in the Test Tube." *The New York Times,* December 13, 1991.

Preston, Richard (1994). *The Hot Zone.* New York: Random House. About the 1989 Ebola-type virus outbreak in Reston, Virginia. Well written.

Rogers, Naomi (1992). *Dirt and Disease: Polio Before FDR.* New Brunswick, NJ: Rutgers Univ. Press. Thorough and clearly written.

Schmaljohn, C. S., et al. (1985). "Antigenic and Genetic Properties of Viruses Linked to Hemorrhagic Fever with Renal Syndrome." *Science,* 227, pp. 1041–1044.

Siebert, Charles (1994). "Smallpox Is Dead, Long Live Smallpox." *The New York Times Magazine,* Aug. 21, pp. 31ff.

Stanley, N. F. (1980). "Man's Role in Changing Patterns of Arbovirus Infections." In Stanley and Joske (see Core Bibliography), pp. 152–173.

Sundin, Daniel, et al. (1987). "A G1 Glycoprotein Epitope of La Crosse Virus." *Science,* 235, pp. 591–593.

Walsh, John (1988). "Rift Valley Fever Rears Its Head." *Science,* 240, pp. 1397–1399.

CHAPTER ELEVEN

Relevant Core Bibliography references are in Kiple, Lederberg and Shope, White and Fenner, and especially Morse.

Aggleton, Peter, et al. (1994). "Risking Everything? Risk Behavior, Behavior Change, and AIDS." *Science,* 265, pp. 341–345.

Anderson, Roy, and Robert May (1992). "Understanding the AIDS Pandemic." *Scientific American,* 266(5), pp. 58–66.

Barinaga, Marcia (1992). "Furor at Lyme Disease Conference." *Science,* 256, pp. 1384–1385.

Bartlett, Christopher L., Alastair D. Macrae, and John T. Macfarlane (1986). *Legionella Infections.* London: Edward Arnold.

Bergdoll, Martin, and Joan Chesney (1991). *Toxic Shock Syndrome.* Boca Raton, FL: CRC Press.

Brandt, Allan (1988). See above, chap. 8.

Breiman, Robert, et al. (1990). "Association of Shower Use with Legionnaires' Disease." *Journal of the American Medical Association,* 263(1), pp. 2924–2926.

Burgdorfer, Willy (1993). "Discovery of *Borrelia burgdorferi.*" In Coyle (see below), pp. 3–7. Discusses the earliest reports of Lyme disease.

Busvine, J. R. (1980). "The Evolution and Mutual Adaptation of Insects, Microorganisms, and Man." In Stanley and Joske (see Core Bibliography), pp. 55–68.

Cohen, Jon (1992). "Debate on AIDS Origin: *Rolling Stone* Weighs In." *Science,* 255, p. 1505.

Coyle, Patricia K., ed. (1993). *Lyme Disease.* St. Louis, MO: Mosby.

Desowitz, Robert (1981). *New Guinea Tapeworms and Jewish Grandmothers.* New York: Norton. Charming essays on parasitology.

Gallo, Robert, and Flossie Wong-Staal (1990). *Retrovirus Biology and Human Disease.* New York: Dekker.

Ginsberg, Howard, ed. (1993). *Ecology and Environmental Management of Lyme Disease.* New Brunswick, NJ: Rutgers Univ. Press.

Gregg, Charles (1983). *A Virus of Love.* Albuquerque: Univ. of New Mexico Press. Essays for general readers about medical detection.

Grmek, Mirko (1990). *History of AIDS: Emergence and Origin of a Modern Pandemic* (tr. Russell Maulitz and Jacalyn Duffin). Princeton, NJ: Princeton Univ. Press. Important for scholars.

Harden, Victoria (1990). *Rocky Mountain Spotted Fever: History of a Twentieth-Century Disease.* Baltimore: The Johns Hopkins Univ. Press.

Horton, Tom (1991). "Deer on Your Doorstep." *The New York Times Magazine,* April 28, pp. 29ff.

Merson, Michael (1993). "Slowing the Spread of HIV: Agenda for the 1990s." *Science,* 260, pp. 1266–1268.

"Mist in Grocery's Produce Section Is Linked to Legionnaires' Disease." *The New York Times,* Jan. 11, 1990.

Morse, Stephen (1993). "Examining the Origins of Emerging Viruses." In Morse (see Core Bibliography), pp. 10–28.

Norman, Colin (1986). "Sex and Needles, Not Insects and Pigs, Spread AIDS in Florida Town." *Science,* 234, pp. 415–417.

Palca, Joseph (1992). "Human SIV Infections Suspected." *Science,* 257, p. 606.

Persing, David, et al. (1990). "Detection of *Borrelia burgdorferi* DNA in Museum Specimens of *Ixodes dammini* Ticks." *Science,* 249, pp. 1420–1423.

"Pigs, AIDS and Belle Glade." *The New York Times,* June 3, 1986.

"Rare Virus May Be Spreading." *The New York Times,* Sept. 25, 1990.

Silverstein, Arthur (1981). *Pure Politics and Impure Science: The Swine Flu Affair.* Baltimore: The Johns Hopkins Univ. Press.

Sternberg, Steve (1992). "HIV Comes in Five Family Groups." *Science,* 256, p. 966.

Temin, Howard (1993). "The High Rate of Retrovirus Variation Results in Rapid Evolution." In Morse (see Core Bibliography), pp. 219–225.

"Theory Tying AIDS to Polio Vaccine Is Discounted." *The New York Times,* Oct. 28, 1992.

Thomas, Gordon, and Max Morgan-Witts (1982). *Anatomy of an Epidemic.* Garden City, NY: Doubleday. Account of the 1976 legionellosis outbreak.

"Turning Tick Bites into Dollars." *The New York Times,* June 4, 1989.

Weiss, Robin (1993). "How Does HIV Cause AIDS?" *Science,* 260, pp. 1273–1279.

CHAPTER TWELVE

In the Core Bibliography see Fiennes, Lederberg and Shope, White and Fenner, and a number of articles in Kiple. The subject of viruses, genes, and cancer is, even at its simplest, challenging to nonspecialists. Reports on cancers of the liver and cervix offer the best introduction.

Aldhouse, Peter (1992). "French Officials Panic over Rare Brain Disease Outbreak." *Science,* 258 pp. 1571–1572.

Alvey, Julie (1990). *Genital Warts and Contagious Cancers.* Jefferson, NC: McFarland. A good, educational introduction for general readers.

Aral, Sevi, and King Holmes (1991). Sexually Transmitted Diseases in the AIDS Era." *Scientific American,* 264(2), pp. 62–69. Good summary for nonspecialists.

Beardsley, Tim (1990). "Oravske Kuru." *Scientific American,* 263(2), pp. 24–26.

Björnsson, J., et al., eds. (1994). *Slow Infections of the Central Nervous System.* Annals of the New York Academy of Sciences, vol. 724. New York: The New York Academy of Sciences.

Bloom, Barry, and J. Murray Christopher (1992). *"Tuberculosis: Commentary on a Reemergent Killer."* *Science,* 257, pp. 1055–1064.

Boren, Thomas, et al. (1993). "Attachment of *Helicobacter pylori* to Human Gastric Epithelium Mediated by Blood Group Antigens." *Science,* 262, pp. 1892–1895.

Bowie, William R., et al., eds. (1990). *Chlamydial Infections.* New York: Cambridge Univ. Press.

Brandt, Allan (1987). See above, chap. 1.

Cates, Willard, Jr. (1988). "The 'Other STDs': Do They Really Matter?" *Journal of the American Medical Association,* 259(24), pp. 3606–3608.

Cherfas, Jeremy (1990). "Mad Cow Disease: Uncertainty Reigns." *Science,* 249, pp. 1492–1493.

"Crack and Resurgence of Syphilis Spreading AIDS Among the Poor." *The New York Times,* Aug. 20, 1989.

"A Dangerous Form of Strep Stirs Concern in Resurgence." *The New York Times,* June 8, 1994.

Ewald, Paul (1993). *Evolution of Infectious Disease.* New York: Oxford Univ. Press.

——— (1993). "The Evolution of Virulence." *Scientific American,* 265(4), pp. 86–93.

"Far Away from the Crowded City, Tuberculosis Cases Increase." *The New York Times,* Dec. 6, 1992.

Goldfarb, Lev, et al. (1992). "Fatal Familial Insomnia and Familial Creutzfeldt-Jakob Disease." *Science,* 258, pp. 806–807.

Goodfield, June (1985). *Quest for the Killers.* Boston: Birkhauser. Good accounts for general readers of research on kuru, hepatitis B, and other infections.

Grmek, Mirko (1990). See above, chap. 11.

"Henson Death Shows Danger of Pneumonia." *The New York Times,* May 29, 1990.

Johnson, Howard M., Jeffrey Russell, and Carol Pontzer (1992). "Superantigens in Human Disease." *Scientific American,* 266(4), pp. 92–101.

Koff, Raymond (1994). "Solving the Mysteries of Viral Hepatitis." *Scientific American/ Science & Medicine,* 1(1), pp. 24–33. A good, up-to-date introduction for serious readers.

Kolata, Gina. (1987). "Are the Horrors of Cannibalism Fact or Fiction?" *Smithsonian,* 17, pp. 151–170.

Krause, Richard (1992). "The Origin of Plagues: Old and New." *Science,* 257, pp. 1073–1078. TSS and other infections.

"Mystery of Balanchine's Death Is Solved." *The New York Times,* May 8, 1984.

Parsonnet, J., et al. (1991). "*Helicobacter pylori* Infection and the Risk of Gastric Carcinoma." *The New England Journal of Medicine,* 325, pp. 1127–1131.

Prusiner, Stanley (1991). "Molecular Biology of Prion Disease." *Science,* 252, pp. 1515–1522.

———— (1994). "The Prion Diseases." *Scientific American,* 272(1), pp. 48–57.

"Resistance to Antibiotics" (1994). *Science,* 264, pp. 359–393. A dozen articles on resistance; useful and up-to-date. For specialists and serious readers.

"Russians Sniff at Beef, Miffing British Donors." *The New York Times,* Jan. 7, 1992.

Ryan, Frank (1993). *The Forgotten Plague.* Boston: Little, Brown. On the history of tuberculosis.

"Study Finds TB Danger Even in Low-Risk Groups." *The New York Times,* Oct. 18, 1992.

"U.S. Panel Urges That All Children Be Vaccinated for Hepatitis B." *The New York Times,* March 1, 1991.

"Viral Sexual Diseases Are Found in 1 of 5 in U.S." *The New York Times,* April 1, 1993.

Webster, Linda, and Robert Rolfs (1991). "Surveillance for Primary and Secondary Syphilis—United States, 1991." *MMWR,* 42(SS-3), pp. 13–19.

Weinberg, Robert (1991). "Tumor Suppressor Genes." *Science,* 254, pp. 1138–1146.

Weiss, Rick (1992). "On the Track of 'Killer' TB." *Science,* 255, pp. 148–150.

Wright, Karen (1990). "Bad News Bacteria." *Science,* 249, pp. 22–24.

CHAPTER THIRTEEN

Relevant core references are Croll and Cross, Fiennes, Lederberg and Shope, McKeown, Morse, Stanley and Joske, White and Fenner, and Wood. See especially the first two chapters in Morse, by Lederberg and Morse respectively.

Anderson, R. M., and R. M. May (1982). See above, chap. 5.

"Army Warns of Diseases Linked to War." *The New York Times,* March 21, 1991.

Beinin, Lazar (1985). *Medical Consequences of Natural Disasters.* Berlin: Springer.

Bongaarts, John (1994). "Can the Growing Human Population Feed Itself?" *Scientific American,* 270(3), pp. 36–42.

"Cambodia's Voiceless." *The New York Times,* May 2, 1991.

Catania, Joseph, et al. (1992). "Prevalence of AIDS-Related Factors and Condom Use in the United States." *Science,* 258, pp. 1101–1106.

"Census Bureau Lifts Population Forecast." *The Wall Street Journal,* Dec. 4, 1992.

Centers for Diease Control. "Viral Agents of Gastroenteritis." *MMWR: Recommendations and Reports,* 39(RR-5).

Cohen, Mitchell (1992). "Epidemiology of Drug Resistance: Implications for a Post-Microbial Era." *Science,* 257, pp. 1050–1055.

————, and Robert Tauxe (1986). "Drug-Resistant *Salmonella* in the United States." *Science,* 234, pp. 964–969.

"Death Toll from Cholera Rising in South Yemen City Hit by War." *The New York Times,* Sept. 24, 1994.

Desowitz, Robert (1981). See above, chap. 11.

———— (1991). *The Malaria Capers.* New York: Norton. Well written. Discusses malaria and related ills, and shows how global campaigns against infection can founder and fail.

"Disease Stalks Iraq." *The New York Times,* June 24, 1991.

"Doctors Say a New Cholera Poses a Worldwide Danger." *The New York Times,* Aug. 13, 1993.

Edison, Millicent, et al. (1988). See above, chap. 2.

Ehrlich, Paul, and Anne Ehrlich (1990). *The Population Explosion.* New York: Simon & Schuster. One of the more dire views of world population growth.

"Epidemic Traced to Feces of Beavers Is Easing." *The New York Times,* Jan. 11, 1986.

"Faster Slaughter Lines Are Contaminating Much U.S. Poultry." *The Wall Street Journal,* Nov. 16, 1990.

Fisher, Jeffrey (1994). *The Plague Makers.* New York: Simon & Schuster. On creating antibiotic resistance in microbes.

"Frontiers in Medicine: Vaccines." *Science,* 265, pp. 1371–1404. A wide-ranging scientific roundup.

Glass, R. I., M. Libel, and A. D. Branding-Bennett (1992). "Epidemic Cholera in the Americas." *Science,* 256; pp. 1524–1525.

"Global Change." (1992). See above, chap. 1.

"Global Warming Threatens to Undo Decades of Conservation Efforts." *The New York Times,* Feb. 25, 1992.

"Illness That Killed 63 in France Is Traced to Pork." *The New York Times,* Feb. 28, 1993. On listeriosis.

Kates, Robert (1994). "Sustaining Life on Earth." *Scientific American,* 271(4), pp. 114–122.

"Kazakhstan Authorities Reported a Sharp Rise in Cholera." *The Wall Street Journal,* Sept. 14, 1993.

Keyfitz, Nathan (1994). "Demographic Discord." *The Sciences,* 34(5), pp. 21–27. Interesting essay on scientific dispute over population and the environment.

Kilbourne, Edwin (1987). See above, chap. 7.

———— (1993). "Afterword: A Personal Summary." In Morse (see Core Bibliography), pp. 290–295.

Livi-Bacci, Massimo (1992). *A Concise History of World Population* (tr. Carl Ipsen). London: Blackwell. A thoughtful view of world demographics; rather hopeful.

Lovejoy, Thomas (1993). "Global Change and Epidemiology: Nasty Synergies." In Morse (see Core Bibliography), pp. 261–268.

"Many Children Are Getting Whooping Cough." *The New York Times,* Dec. 17, 1993.

Mims, Cedric (1980). See above, chap. 3.

Mitchison, Avrion (1993). See above, chap. 1.

Neu, Harold (1992). "The Crisis in Antibiotic Resistance." *Science,* 257, pp. 1064–1072.

Nowak, Rachel (1994). "Flesh-Eating Bacteria." *Science,* 264, p. 1665.

"Outbreak of Disease in Milwaukee Undercuts Confidence in Water." *The New York Times,* April 20, 1993.

Parrish, Colin (1993). "Canine Parvovirus 2: A Probable Example of Interspecies Transfer." In Morse (see Core Bibliography), pp. 194–202.

"Personal Health: Why the Food You Eat May Be Hazardous to Your Health." *The New York Times*, Oct. 5, 1994.

Peters, Robert, and Thomas Lovejoy (1992). *Global Warming and Biological Diversity.* New Haven, CT: Yale Univ. Press.

Pickering, Larry (1986). "The Day Care Center Diarrhea Dilemma." *American Journal of Public Health,* 76(2), pp. 623–624.

"Population Alarm" (1992). *Science,* 255, p. 1358.

"Protozoon Makes Bid to Move into the Scientific Mainstream: Cryptosporidium" (1990). *The Scientist,* 4(21), pp. 4–7.

"Researchers Study How Environment May Influence the Spread of Cholera." *The Wall Street Journal,* July 10, 1992.

Roberts, Leslie (1988). "Is There Life After Climate Change?" *Science,* 242, pp. 1010–1012.

"Russia Fights a Rising Tide of Infection." *The New York Times*, Oct. 2, 1994.

"The Rwandan Disaster: The Stark Assessments." *The New York Times,* July 22, 1994.

Seaman, J., S. Leivesley, and C. Hogg (1984). *Epidemiology of Natural Disasters.* Basel: Karger.

Sen, Amartya (1993). "The Economics of Life and Death." *Scientific American,* 268(5), pp. 40–47.

Service, Robert F. (1994). "*E. coli* Scare Spawns Therapy Search." *Science,* 265, p. 475.

Simil, Vaclav (1994). *Global Ecology.* London: Routledge.

Sternberg, Steve (1994). "The Emerging Fungal Threat." *Science,* 266, pp. 1632–1634.

Stix, Gary (1993). "Red-Banner Burger: Toward Food Inspection That Assures Safety." *Scientific American,* 268(12), pp. 132–135.

"Stoked by Ethnic Conflict, Refugee Numbers Swell." *The New York Times*, Nov. 10, 1993.

"Study of Retail Fish Markets Finds Wide Contamination and Mislabeling." *The New York Times,* Jan. 16, 1992.

"Warning: Venturing into Ex-U.S.S.R. May Well be Hazardous to Your Health." *The Wall Street Journal,* March 26, 1993.

Wilson, E. O. (1992). *The Diversity of Life.* Cambridge, MA: Harvard Univ. Press.

index

Aborigines, Australian, 110
Abu Hureyra. *See* Tell Abu Hureyra
Acetylsalicylic acid, 140
Adaptation, host-parasite, 17–18
Adult diarrhea, 219
Adult T-cell leukemia (ATL), 6, 188
Aedes aegypti, 106, 158–59, 160,
    163–64
Aedes albopictus, 160
Afghanistan, 132, 217
Africa, 31, 40, 42, 74, 94, 106, 157,
    192, 202, 220
  AIDS in, 185
  cholera epidemic, 215–16
  dengue in, 159
  epidemics from, 70
  leprosy in, 82
  malaria in, 41
  new diseases, 163–67
  polio in, 154
  retroviruses from, 188
  syphilis in, 123, 127
  tuberculosis in, 210
  yellow fever epidemics, 159
Aging of population, 3, 224
Agricultural Revolution, 29–30
  twentieth century, 155
Agriculture, 34–36, 98, 140–41
  and disease, 31, 38–42
  and population increase, 41
  primitive, in Americas, 97, 98

AIDS, 2, 4, 6, 9–10, 175, 185–94,
    218, 224
  conspiracy theories, 135
  origin, 11, 39
  polio and, 154
  spread, 225
Air conditioners, 183–84, 223
Airborne diseases, 52
  coccidioidomycosis, 217
Airline employee, spread of AIDS, 192
Airplanes, and spread of disease, 7,
    222–23
Alexandria, Egypt, 82
Alzheimer's disease, 5, 200
Amazonia, ecological disruption, 162
Amebiasis, 206–7
American Museum of Natural History
    (New York City), 150–51
Americas (New World), 91, 96–97
  cultural development, 97
  mass extinctions, 31
  syphilis in, 123, 125
  typhus introduced into, 115
  See also South America; United States
America's Forgotten Pandemic (Crosby),
    145
Amerindians, 58, 96–106
Anasazi, 97
Anemias, hereditary, 40
Animal husbandry, 34
Anisakiasis, 220

Anopheles gambiae, 40
Anthrax, 24, 37, 51, 138
Antibiotics, 139, 155, 180, 205, 211
   resistance to, 206, 212-13, 227
   and tuberculosis, 208, 209
Antiquity, diseases of, 55-56
Antisepsis, internal, 139
Anti-STD programs, 207-8
Antitoxins, 139
   diphtheria, 146
Antoninus, plague of, 70, 71
Arab culture, fifteenth century, 94
Arab world, cholera in, 132
Arboviruses (arthropod-borne
      infections), 54, 156-57
   encephalitis, 157-58
   hantaviruses, 169-73
   hemorrhagic fevers, 158-69
Arenavirus 3080, 166
Aretaeus, 146
Argentina, 5, 161, 216
Argentine hemorrhagic fever,
      6, 161
Arthritis, 19
   from Lyme disease, 178, 179
   in Neolithic farmers, 35, 98
Arthropods, 54, 156. *See also*
      Arboviruses
Asia, 31, 70, 86, 94, 157,
      169, 220
   AIDS in, 185
   bubonic plague in, 90
   cholera in, 215-16
   dengue in, 159, 160
   influenza pandemics, 143, 144
   tuberculosis in, 210
Asian flu, 143
Asian tiger mosquito, 160
Aspirin, 140
Aswan High Dam, Egypt, 163
Athens, great plague, 59-60
ATL (adult T-cell leukemia), 188
Austin, Texas, legionellosis in, 183
Australia, 25, 110, 123, 157-58,
      159, 180
Australopithecines, 21
Autoimmune diseases, 188
Avignon, bubonic plague in, 89
Azore Islands, 95, 96
Aztecs, 97
   disease epidemics, 101-2, 103

Babesiosis, 6, 7, 180-81, 220
Bacteria, 14, 16
   beneficial, 17
Bacterial infections, 49
Bacteriophages, 212
Bahama Islands, 103
Balanchine, George, 195-96
Balkans, 89, 169
Baltimore rat virus (BRV), 170
Balto (Alaskan husky), 145-47
Bangladesh, cholera in, 215, 217
Bartonellosis, 100
BCG vaccine, 208
Bear meat, raw, 23
Beaver feces, and giardiasis, 219
Bedbugs, 112
Bede, St., 71
Behavioral change, 225
   and diseases, 18, 21-23, 205,
      229-30
   AIDS, 185, 190, 193
   genital herpes, 203
   new, 1, 175-76
   oncoviruses, 201
   sexually transmitted, 207
Bejel, 126-27
Belle Glade, Florida, AIDS in, 186
Beringia, 97
Bhaibulaya, Manoon, 45
Bible, accounts of epidemics in, 55-56
Biosafety Level 4 germs, 161-62
Biosocial change, 5
Biosphere, 4
   changes in, 11
Birds, 36, 38, 156, 158
Birth control movement, 67
Bitterroot Valley, Montana, 177
Black Assizes, 115-16
Black Death, 86, 90. *See also* Bubonic
      plague
Black rats, 76-77, 87. *See also* Rat(s)
Blacks, repatriation from Americas to
      Africa, 108
Blaithmac (Irish king), 81
Blindness, preventable, 206
Blood transfusions, 223, 224
   and hepatitis B virus, 202-3
Boccaccio, Giovanni, 88
Body lice, 113-14
Bogalusa, Louisiana, legionellosis, 184
Bolivia, 161-62, 216

Bolivian hemorrhagic fever, 6, 162
Borrelia burgdorferi, 177-79
Boston, 106-7, 145
Botulism, 24
Boule, Pierre, 26
Bovine spongy encephalopathy (BSE),
    198-99
Bovine tuberculosis, 38
Bradford, William, 104
Brain diseases, slow viruses and,
    198-99
Brain-tissue grafts, and Creutzfeldt-
    Jakob disease, 200
Brazil, 5, 154, 157, 216
Brazilian purpuric fever, 6, 227
British colonization of India, 132
Broad-spectrum revolution, 30, 32, 33
Bronze Age, 48, 50, 69
    diseases, 51, 54, 62
Brucellosis, 19, 24, 38, 53
BRV (Baltimore rat virus), 170
Bryan, William Jennings, 30
BSE (bovine spongy encephalopathy),
    198-99
Bubonic plague, 4, 24, 39, 48,
    52, 74-78, 79, 86-91, 94,
    135, 222
Bulinus snail, 62
Burgdorfer, Willy, 177, 178-79
Burkitt's lymphoma, 201
Burnet, Macfarlane, 13-14
Bushmen, 25
Byzantine empire, 75, 89

Calcutta, 123, 130, 132
California, 172, 217
Calomys callosus, 161
Calomys musculinus, 161
Campagna, Italy, malaria in, 70
Campylobacter jejuni, 219
Canary Islands, 95-96
Cancer, 5, 9, 201-2, 205
    treatments for, 224
Candida fungi, and AIDS, 224
Cannibalism, ritual, 198
Capillaria philippinensis, 44-45
Carbohydrates, and disease, 35
Carcinogens, 4
Caribbean islands, dengue
        epidemic, 160
Carnivores, 22

Carriers of diseases. See Transmission
    of diseases
Carrión's disease, 100
Carthage, ancient, epidemic in, 66
Cats, domestic, 36, 37, 199
Cattle, and disease, 39
Cave bears, arthritis of, 19
CDC. See Centers for Disease Control
Cells
    evolution, 14
    metabolism, slow viruses and, 200
Cellular biology, 138
Centers for Disease Control (CDC), 8
    and AIDS, 186
    and dengue, 160
    and Lassa fever, 165
    and legionellosis, 183, 184
    and salmonellosis, 220
    and tick-borne diseases, 177
    and tuberculosis, 210
Central America, epidemics, 103
Central Park (New York City), 145-47
Cervical cancer, 6, 9, 203-5
Chadwick, Edwin, 137
Chagas' disease, 98, 100, 112
Chancroid, 206, 207
Charles I, King of Spain, 115
Charles VIII, King of France, 123
Charleston, South Carolina, yellow
        fever epidemic, 107
Chicago, Lassa fever threat, 165
Chickenpox, 10, 58, 59, 203
Chikungunya, 6, 168
Childbirth fever, 139
Childhood anemia, 35
Childhood diarrhea, 219
Childhood diseases, endemic, 18, 109
Children, 218
    Neolithic, 38
Chile, cholera pandemic, 216
China, 70, 73-74, 86, 94, 169, 225
    bubonic plague in, 75-76, 87, 90
    cholera in, 132
    dengue in, 159
    leprosy in, 82
    Lyme disease in, 180
    polio in, 154
    syphilis in, 123
Chipmunks, and bubonic plague, 24
Chlamydia, 4, 6, 15, 206
Chloroplasts, 15

Cholera, 2, 6, 42, 53, 111, 120, 121,
    128, 129, 130–38, 215
  in Australia, 110
  in former Soviet Union, 218
Christianity
  and lice, 113
  and pandemics, 66
Chronic diseases, 19, 20, 48
Chronic fatigue syndrome, 5
Church, Roman Catholic, bubonic
    plague and, 90
Cincinnati, new hantaviruses in, 170
Cirrhosis, hepatitis and, 202, 203
Cities, 47–48, 80, 130, 136, 142, 225
  in ancient Americas, 97
  and disease, 48–63
  U.S., new hantaviruses in, 170
Civil war, and disease, 218
Civilization, 30
  diseases of, 3–4
CJD. *See* Creutzfeldt-Jakob disease
Clement VI, Pope, 88, 89
Clement VII, Pope, 115
Climate
  and disease, 25, 62, 73–74
  Middle Ages, 79–80, 86
Climax state, ecological, 67–68, 80
Clothing, and disease, 52–53, 111–12
Cockburn, Aidan, 131
Codworm, 220
Cold(s), 11, 48, 52
Cold sores, 20
Cold War, end of, 218
Colombia, cholera pandemic, 216
Colonialization, European, 96
Colorado, hantavirus pulmonary
    syndrome in, 172
Columbus, Christopher, 93–94, 96
Commensalism, 18
Complex cells, origin of, 14–15
Cone-nose bug, 98, 112
Conjunctivitis, 226
Connecticut, babesiosis in, 181
Conspiracy theories for diseases,
    135, 186
Consumption, 84–85. *See also*
    Tuberculosis
Contact, diseases spread by, 52–53
Contagion theory, 88, 133–34, 138–39
Contraception, 142
Cook, James, 110

Copulation, and spread of disease, 122
Cornea transplants, and Creutzfeldt-
    Jakob disease, 200
Cortés, Hernán, 101
Costa Rica, dengue in, 159
Crack-cocaine, and syphilis, 206
Creation, supposed date, 30
Creutzfeldt-Jakob disease (CJD), 196,
    198, 200, 223
Crimean-Congo hemorrhagic fever, 168
Crops, primitive, in Americas, 97
Crosby, Alfred, 50, 96, 145
Cross, John, 45
Crowd diseases (zymotics), 48–49,
    60–78, 130, 139
  American colonists and, 101–10
  Amerindians and, 100–101
  in ancient cities, 52, 56–57
  Australian aborigines and, 110
  immunities to, 141
Crusades, 80, 85, 94
Cryptococcus infections, and AIDS, 224
Cryptosporidium, 219
Cuba, epidemics, 102, 106, 133, 160
Culture
  changes in, 111, 126, 128
  diet and, 22
  prehistoric, 30
Cyprian, St., 66, 71
Cytomegalovirus, and AIDS, 224

Da Gama, Vasco, 96, 123
Da Ponte, Lorenzo, 133
Dark Ages, 78, 79–80, 85
Darwin, Charles, 98
Day care, children in, 224
De Chauliac, Guy, 88
De Kruif, Paul, 139
De Soto, Hernando, 102
Deafness, of Neolithic fishermen, 98
Death, causes of, 48
  bubonic plague, 89
  measles, 58
Decameron (Boccaccio), 88
Deer population, eastern U.S., 179–80
Deer ticks, 176–77
Deficiency diseases, 51
Defoe, Daniel, 90
Degenerative neurological diseases,
    200–201
Demographics, changes in, 224–25

Dengue, 40, 52, 54, 158, 159, 160
Dengue hemorrhagic fever (DHF), 6, 159–60
Dengue shock syndrome (DSS), 10, 160
Denmark, syphilis in, 123
Dental disease, 19, 35, 98
Depression, 201
Dermot (Irish king), 81
Developed nations, 7, 225
Developing nations, 43, 207
DHF (dengue hemorrhagic fever), 6, 159–60
Diagnosis of ancient diseases, 56
Diama Dam, and Rift Valley fever, 163
Diarrheal diseases, 43, 53, 131, 139, 219, 223, 229
    sanitation and, 138
Diaz de Isla, 125
Dickson Mound burials, 35–36, 99
Die-offs, mass, 65, 68–69, 105
Diet, 21–22, 98, 140–41
    in cities, 50–51
    and disease, 35, 45
        prehistoric, 32–33
    of Mayans, 99
Dietary laws, 54
Dinosaurs, infections of, 14
Diphtheria, 2, 5, 6, 14, 53, 139, 141, 146, 218
    in New World, 103, 108
Disease(s), 10, 20–23
    of Amerindians, 98
    causes of emergence, 217–18
    cities and, 48–63
    European colonization and, 100–110
    extinction of, 81
    famine and, 86
    new, 1–11, 13. See also New diseases
    severity of, 16
    See also Epidemics
Disease-resistant crops, 118
Distemper in dogs, and measles, 38
DNA, 14, 187
Doctors
    and cholera, 133–34
    and hepatitis B virus, 202–3
Dogs, 36, 37, 39, 57
Dolphins, 122
Domestic animals, 36–39, 143

Dourine, 122
Drug-induced immuno-
    suppression, 224
Drug resistance, 4, 210
    changes in, 212–13
DSS (dengue shock syndrome), 160
Dwarfism, hormone treatment for, 200
Dysentery, 42, 100, 107, 119, 217, 218
    sanitation movement and, 137–38

East Africa, cholera in, 132
Eastern equine encephalitis (EEE), 157, 160
Ebola fever, 6, 8, 166–67
EBV (Epstein-Barr virus), 201
Echinococcosis, 36, 51
Ecological change, 18, 54, 161–63, 176
    and Lassa fever, 164
    and Lyme disease, 179–80
Ecological imperialism, 96
Ecology, 67, 229
Economic development, 220
    and diseases, 221
Economy, European, plague and, 90
Ectoparasites, 111–14
Ecuador, cholera pandemic, 216
Education
    and population growth, 225
    and spread of AIDS, 193
    about STDs, 207–8
EEE (eastern equine encephalitis), 157, 160
Egypt, 82
    Rift Valley fever epidemic, 163
Ehrlich, Paul, 139
Ehrlichiosis, 177
Eighteenth-century influenza pandemics, 143
El Tor strain of cholera, 215–17
Elderly, in group homes, 224
Elephantiasis, 37
Elizabeth I, Queen of England, 109
Emigration, nineteenth-century Ireland, 120
Empires, ancient Americas, 97
Enamel hypoplasia, 33
Encephalitis, 7, 24, 41, 49, 157–58
    Asian tiger mosquito and, 160
    in Bronze Age cities, 52

Endemic diseases, 18, 62, 110
  dengue, 159
  Ebola fever, 166
  syphilis, 126
  typhus, 42, 113-14
Enders, John, 153
England, 80, 84, 112, 123, 124
  disease epidemics, 81
    bubonic plague, 88, 89
    cholera, 132-33, 135-37
    Great Plague of 1665, 90
    typhus, 115-16
  and Irish Potato Famine, 119-20
  nineteenth-century medicine, 134
  smallpox inoculation, 141-42
English Reformation, typhus and, 115
English settlers in New World, 104
Environment, 67, 229
  Neanderthalers and, 28
Environmental change, 3, 4
  and disease, 11, 18, 20, 39-40,
    156, 229-30
    AIDS, 185, 190
    encephalitis viruses, 157
    epidemics, Middle Ages, 86
    legionellosis, 181-84
    Lyme disease, 179-80
    new diseases, 1, 2, 161-63,
      175-76, 220
    hunter-gatherer societies and, 31-32
Epidemic diseases, new, 1-11
Epidemics, 17-18, 193-94
  Canary Islands, 95
  cities and, 48-63
  global, 3
  morality and, 66
  in New World, 101-3
  Pacific islands, 110
  population and, 68, 130, 226
  prehistoric, 19-20
  See also Pandemics
Epidemiology, 3-4
Epstein-Barr virus (EBV), 201
Eradication of diseases, 154-55
Erysipelas, 58
Erythema migrans, and Lyme disease,
  178-79
Escherichia coli, 17, 226
  new strain, 2, 6, 220
Essay on the Principle of Population, An
  (Malthus), 66-67

Ethiopia, 217
Ethnic customs, 54
Euglena, 15
Eurasia, encephalitis in, 157-58
Europe, 31, 80-81, 86, 94
  diseases in
    AIDS, 185
    babesiosis, 181
    bubonic plague, 87, 89-90
    cholera, 132-33
    encephalitis, 158
    hepatitis B, 202
    influenza, 143
    leprosy, 82
    Lyme disease, 180
    malaria, 41
    smallpox, 101
    syphilis, 127-28
    tuberculosis, 210
    typhus, 115-16
  epidemics in, 73, 84
  post-plague expansion, 94-95
European colonies, 96
  and slavery, 105-6
Evolution, 14-15, 30
  of diseases, 13-14
Exanthemata, 58
Exploration, 93-96, 109, 110
Extinctions, 19, 31
  of diseases, 81, 155

Famine, 51, 86, 118-21, 225, 217
Famine fever, 119, 120
Farm animals, and influenza, 143
Farm productivity, increase of, 155
Farr, William, 137, 138
Fatal familial insomnia (FFI), 200
Feline immunodeficiency virus
  (FIV), 187
Female sterility, preventable, 206
Fertilizers, 41, 155
Feudalism, plague and, 90
FFI (fatal familial insomnia), 200
Fiennes, Richard, 82
Fifteenth century, 94-95
Fiji, measles epidemic, 58-59, 110
Filarial worms, 37
Filtration of water systems, 219
Finland, Puumala virus, 169
Fish, undercooked, 45-46, 220
Fish tapeworm, 33

Fishing, diseases from, 98
FIV (feline immunodeficiency
    virus), 187
Flagellants, 88
Fleas, 112, 113–14
Flesh-eating infections, 9
Flexner, Simon, 151
Florida, 7, 160, 172
Flu. *See* Influenza
Fluoridation, 3
Food-borne infections, 218–19,
    220, 223
Food chain, 16
Food poisoning, 53
Food production, and population
    growth, 50
Fore tribe, New Guinea, 197–98
Fossil record, 14, 19, 21–22
Four Corners virus, 2, 10,
    170–73, 217
Fracastoro, Girolamo (Fracastorius),
    115, 124–25
France, 81, 115, 123, 124, 133–34
    epidemics, 72
        bubonic plague, 89, 90
French Revolution, typhus and, 116
French settlers in New World, 104
Fungi, 16
    potato blight, 118–21

Gajdusek, D. Carleton, 197–98, 200
Galen, 58, 62
    plague of, 71
Game, wild, diseases from, 24–25
Gangrene, 19, 24
"Garden of Germs, The," exhibit,
    150–51
Genetic exchanges in bacteria, 212
Genital herpes (HSV 2), 4, 6, 9, 122,
    203–4
Genital warts, 9, 204
Germ theory, 138–39, 142
German measles, 58
Germany, 81, 112, 123, 124,
    132, 180
Gerstmann-Sträussler syndrome
    (GSS), 200
Giardia, and AIDS, 224
Giardiasis, 6, 206–7, 219
Gibbon, Edward, 66
Glanders, 51

Global die-off, 226
Global warming, 216–17, 221–22
Goats, 37, 39
Gonorrhea, 48, 53, 80–81, 122–23,
    205, 206
Governments, and health, 142
    nineteenth century, 121, 137–38
Granada, siege of, 114–15
Great Dying, 90. *See also* Bubonic
    plague
Great Hunger, Ireland, 118–21, 140
Great pox (syphilis), 124. *See also*
    Syphilis
Greece, 123
    ancient, 50, 51, 62, 84–85
Greenland, bubonic plague in, 91
Gregg, Charles, 24
Gregory of Tours, 72
GRID (gay-related immune disease),
    185. *See also* AIDS
Ground squirrels, 24
GSS (Gerstmann-Sträussler
    syndrome), 200
Guanarito virus, 163
Guanches, 95–96
Guatemala, cholera pandemic, 216
Guillain-Barré syndrome, 8
Guinea pigs, domestication of, 36

Hackett, C. J., 126–27, 128
Haemophilus influenzae, 145,
    226–27
Hairy-cell leukemia, 188
Haiti, 159, 185
Hamburger, undercooked, 220
Hansen's disease, 83. *See also*
    Leprosy
Hantaan virus, 169
Hantavirus pulmonary syndrome
    (HPS), 6, 172–73
Hantaviruses, 2, 10, 169–73, 217
Harris lines, 33
Hausen, Harald zur, 204
Hawaii, measles epidemic, 59, 110
HBV (hepatitis B virus), 201–2
Head lice, 113–14
Health, 13, 50
    of prehistoric peoples, 33
Heart damage, from Lyme disease, 178
Heirloom infections, herpes, 122
Heirloom parasites, 37, 112–13

Helicobacter pylori, 205
Helminths, 37, 42, 43, 44, 220
Hemophilia, and AIDS, 185
Hemorrhagic fever with renal syndrome (HFRS), 169–70
Hemorrhagic fevers, 5–7, 10, 11, 24, 39, 157, 158–59, 168, 220
  in Bronze Age cities, 52
  dengue, 159–60
  Ebola fever, 8
  new, 161–63, 173
  yellow fever, 158
Henry VII, King of England, 81
Henry VIII, King of England, 115
Henson, Jim, 9, 211
Hepatitis, 6, 10, 20, 53, 217, 218
  medical technology and, 223
  types, 202–3
Hepatitis A, 100
Hepatitis B (HBV), 201–2
Hepatitis C, 203
Herbicides, 155
Herbivores, 22
Herbularios, 43–44
Herd diseases, 48, 49
Herpes, 122
  and HIV, 207, 224
  See also Genital herpes; Oral herpes
Herpes simplex viruses, 20, 203
Herpes zoster, 203
Herring worm, 220
Heterosexual transmission of hepatitis B virus, 202–3
HFRS (hemorrhagic fever with renal syndrome), 169–70
Hippocrates, 58, 62, 85
History of the Decline and Fall of the Roman Empire, The (Gibbon), 66
Hittites, plague among, 61
HIV (human immunodeficiency virus), 186–93, 199–200, 207
  and cancer, 201
  medical technology and, 223–24
  and tuberculosis, 209, 210
Holland, syphilis in, 123, 124
Homeless people, and multi-drug-resistant tuberculosis, 209–10
Homo erectus, 21–26
Homo sapiens, 31
Honduras, disease epidemics, 103

Hookworms, 37, 38
Horses, 37, 39
Hospitals, 134, 139, 212–13
  infections in, 211, 223–24
Houston, 170, 210
HPS (hantavirus pulmonary syndrome), 172–73
HPV (human papilloma virus), 9, 204
HTLV (human T-cell lymphotropic virus), 188, 201
  medical technology and, 223
Hudson, Ellis, 126–27, 128
Hull, Thomas, 39
Human(s), prehistoric, 20–22
Human behavior, and evolution of germs, 122, 128
Human disease, reasons for, 18
Human dwellings, and ectoparasites, 112
Human ehrlichiosis, 6
Human immunodeficiency virus (HIV). *See* HIV
Human nature, arguments about, 21–22
Human papilloma virus (HPV), 9, 204
Human pathogens, mutation of, 141
Human T-cell lymphotropic virus (HTLV), 188, 201
Human toxoplasmosis, 6
Hun(s), 71, 72–73
Hundred Years War, 86, 94
Hungary, diseases in, 115, 123, 135
Hunter-gatherer societies, 24–26, 31–32
Hygiene, breakdowns of, 217–18
Hypertension, hantaviruses and, 170
Hypodermic needles, and disease, 191

Iceland, sheep disease in, 196–97
Iditarod dogsled race, 146
Ilocanos, 43–46
Immigrants, to U.S., and polio, 152
Immune system, weakened, 224
Immunities to infections, 49, 141
Incas, 97, 102, 103
India, 73, 74, 132
  bubonic plague in, 78, 87, 90
  cholera in, 130, 215, 217
  hemorrhagic fever in, 163
  leprosy in, 82

India (*continued*)
  new hantaviruses in, 170
  polio in, 154
Indonesia, cholera pandemic, 215
Industrial Revolution, 128, 130, 208
Industrialization, 221, 223
Infant deaths, 3, 140
Infantile paralysis, 149. *See also*
  Poliomyelitis
Infections, 13–28, 48–49
  new avenues of, 4
  regional, 60
  See also Disease(s)
Influenza, 3, 8, 38, 48, 143–46, 229
  adaptation of, 226
  cause, 140
  in New World, 103, 108
  in South Pacific, 110
Inherited diseases, 20
Inoculation, 141–42
Insects, diseases carried by, 24, 54
Institutional settings, infections in, 224
Interdependence, ecological, 67
Interferon alpha, 204
Intestinal infections, 42
  polio virus, 153
  shigellosis, 53
Intestinal parasites, 20, 25, 37–38
  Capillaria, 6, 44–46
  domestic animals and, 36, 37
  in precontact Amerindians, 100
Intravenous drug use, 188, 205
  and AIDS, 185, 191
  and hepatitis B virus, 202–3
Invasive medical procedures, 223–24
Ireland, 133
  famine, 118–21, 140
Iron Age, 50, 62
Iron-deficiency anemia, 35, 99
Irrigation, 41–42, 62, 157, 220
Irritable bowel syndrome, 219
Islam, bubonic plague and, 75, 89
Italy, diseases in, 89, 124
Ixodes scapularis (deer tick), 176–77

Jamestown colony, 107
Janibeg, Khan of Kipchak Tatars, 87
Japan, 42, 44–45, 73, 123, 124,
  132, 180
Japanese encephalitis, 157
Jenner, Edward, 142

Jeon, Kwang, 18
Jessore, cholera in, 132
Jews, blamed for plague, 88, 91
Johns Hopkins Hospital (Baltimore), rat
  virus infections, 170
Johnson, Karl, 166, 228
Journal of the Plague Year, A
  (Defoe), 90
Junin virus, 5, 161
Justinian, plague of, 74–78
Juvenile rheumatoid arthritis
  (JRA), 176

Kaffa, bubonic plague in, 87
Kalahari Desert, 25
Kaposi's sarcoma, 185
Kasson, George, 146–47
Kates, Robert, 225–26
Kidney disease, hantavirus and, 170
Kinshasa, and Ebola fever, 166
Kissing bug, 98
Kliks, Michael, 37
Koch, Robert, 138–40
Korean hemorrhagic fever, 6, 10, 169
Kuru, 197–98, 200
Kyasanur Forest disease, 5, 6, 163

Labor, specialized, and health, 51
Laboratory animals, 8, 164, 167, 191,
  222–23
LaCrosse encephalitis, 6, 7, 122, 157
Land, clearing of, 39–40, 190
Land use, changing, 220, 221
Lascaux cave, 30
Lassa fever, 6, 7, 165–66
Latin America, diseases in, 158, 185
  cholera pandemic, 216–17
Lazar houses, 84
Lederberg, Joshua, 228
Legionellosis (Legionnaires' disease),
  6, 7, 8–9, 175, 181–84, 223
Legionnaires' disease. *See*
  Legionellosis
Lenin, Vladimir I., 117
Lentiviruses (slow viruses), 197–200
Leprosy, 82–84, 110, 139, 229
  and tuberculosis, 85
Leptospirosis, 24, 52
Leroi-Gourhan, Arlette, 27–28
Leukemia, retroviruses and, 201
Lice, 112–14

Life expectancy, 36
    increase in, 1, 3, 140
    of precontact Amerindians, 100
Lifestyle changes, 1, 2
Lima, 216
Lister, Joseph, 140
Listeria, 220
Little Ice Age, 86, 94, 113
Liver cancer, 202, 203
Livestock, 37
Living standard, and population
    growth, 225
Loggers, and yellow fever, 159
London, 89, 117, 133, 136–37
Long Island, New York, diseases,
    177, 181
Lorenz, Konrad, 36
Louis XV, King of France, 109
Louisiana, hantavirus pulmonary
    syndrome in, 172
Louisiana Territory, 133–34
L'Ouverture, Toussaint, 133
Lucayans, 103
Luis I, King of Spain, 109
Lung fluke (*Paragonimus*), 33
Lyme disease, 4, 6, 7–8, 39, 54, 175,
    176–80, 220

Machupo virus, 162
McKeown, Thomas, 140, 141
McLennan, John, 26
McNeill, William, 52, 58, 141
Mad cow disease, 198–99
Madeira Islands, 95, 96
Magellan, Ferdinand, 94, 96
Maize, 99
Malaria, 4, 6, 20, 24, 38, 40–42,
    43, 62, 81, 107, 139, 140,
    217, 229
    in Bronze Age cities, 52
    in China, 74
    in New World, 107, 108–9
    in precontact Amerindians, 100
    in Roman empire, 69–70
Male homosexuals, 185, 188, 205–6,
    207, 219
    and hepatitis B virus, 202–3
Malignant subtertian malaria, 40
Malignant tertian malaria, 54,
    106, 107
Mallon, Mary ("Typhoid Mary"), 152

Malthus, Thomas, 66–67
Manson, Patrick, 139, 151
Maoris, 110
Marburg disease, 6, 7, 164
March of Dimes, 150
Marcus Aurelius, 71
Margulis, Lynn, 15
Marseilles, bubonic plague in, 90
Martha's Vineyard, babesiosis on, 181
Mary II, Queen of England, 109
Mastoid infections, 98
Mather, Cotton, 106–7
Mather, Increase, 104
Matlazahuatl, 103
Mauretania, Rift Valley fever in, 163
Maximilian I, Holy Roman
    Emperor, 123
Maximilian II, Holy Roman
    Emperor, 115
Mayans, 38, 97, 99, 102
Measles, 2, 4, 5, 6, 38, 48, 57–59,
    102–3, 108, 110, 218
Meat-eating, and evolution, 21–22
Mechanized food processing, 223
Medical establishment, and cholera,
    133–34
"Medical student syndrome," 5
Medicine, 134, 142
    advances in, 139–40
    and AIDS, 185
    technology, 205
    and new diseases, 223–24
    and spread of AIDS, 191
Mediterranean area, epidemics, 75
Memphis, Tennessee, yellow fever
    epidemic, 108
Meningitis, 52, 226
Merezhkovsky, Konstantin, 14–15
Mesolithic Age, 33–34, 36
Mesopotamia, endemic diseases, 62
Mexico, 103, 104, 133, 216
Miami, tuberculosis in, 210
Mice, 24, 39, 161, 162, 164, 187
    and hantaviruses, 169, 171–72
    and Lyme disease, 176–77
Microbe Hunters (de Kruif), 139
Microbial drug resistance, 210
Middle Ages, 79–80, 85, 113
Middle East, 73, 87
    ancient, crowd diseases in, 49, 61
    See also Near East

Migrations, European, 109–10
Military, and smallpox inoculation, 142
Milwaukee, diarrhea epidemic, 219
Mindanao, intestinal capillariasis in, 44, 45
Mites, 42
Mitochondria, 15
Mobile Bay (Gulf of Mexico), cholera infection, 216
Molecular biology, 15, 188, 200
Mongol empire, plague and, 90
Monkey(s), 158
    and HIV, 190–91
    See also Laboratory animals
"Monkey disease," 163
Monkeypox, 168–69, 190
Mononucleosis, 122, 201
Montezuma, 101
Morality, epidemics and, 133
    pandemics, 65–66, 88
Morse, Stephen, 193
Moscow, 116, 132
Mosquitoes, 39–40, 156
    diseases carried by, 122, 160–61, 168
    dengue, 10, 159
    encephalitis, 7, 158
    Rift Valley fever, 163–64
    yellow fever, 106, 158–59
Motolinia, Toribio, 101–2
Mound Builders, 97, 98, 102, 105
Mozambique virus, 165–66
Muerto Canyon virus, 173
Multi-drug-resistant (MDR) tuberculosis, 208–10
Multiple infection, of dengue, 160
Multiple sclerosis, 5, 200
Mumps, 38, 48, 57, 103
Muni, Paul, 139
Murray Valley encephalitis, 157–58
Mutations
    of brain protein, 200
    of human pathogens, 141, 226–27
    hantaviruses, 172
    influenza virus, 8, 143, 144
    syphilis bacteria, 127–28
    of RNA viruses, AIDS and, 189–90
Mutualism, 16, 18
Mycenean Greece, decline of, 69
Mycobacteria, 82–85
    tuberculosis, 84, 85, 208–9

Mycoplasma, 15
Mystery Disease of Pudoc, 43–46

Nagana disease, 38
Nandy (Neanderthal man), 26–27
Nantucket Island, babesiosis on, 180–81
Naples, 115, 123
Napoléon Bonaparte, 116, 142
Nasopharyngeal cancer, 201
Native Americans, 58, 96–106
Natural disasters, 43, 86, 217
Natural History of Infectious Diseases (Burnet and White), 13–14
Nature, sentimentality about, 229
Neanderthalers, 19, 26–28, 31
Near East, syphilis in, 123. See also Middle East
Neolithic era, 34, 36, 37–38, 42–43
    environmental change, 39–40
    New World, 98
    population growth, 49–50
Neolithic Revolution, 29–30, 98
Neo-Malthusians, 67
Nephropathia epidemica, 169
Nevada, hantavirus pulmonary syndrome in, 172
New diseases, 1–11, 13, 83, 155–56, 161, 173, 175, 216–30
    cholera as, 131
    hantaviruses, 170–72
    spread of, 227
New Guinea, 127
    Fore tribe, 197–98
    Murray Valley encephalitis, 157
New Mexico, hantavirus, 171–72
New Orleans, 108, 159, 170
New World. See Americas
New York City, 133, 138, 184, 210
    Central Park, 145–47
    hantaviruses, 170
    1918 influenza pandemic, 144
    polio epidemics, 150, 152
    yellow fever epidemic, 107
New York Times, The, 171
New Zealand, European diseases in, 110
Ney, Michel, 116
NGU (nongonococcal urethritis), 206
Nicolle, Charles, 227
Nigeria, Lassa fever in, 7

"Night soil," 41, 42
Nineteenth century, 3, 130, 143
Noguchi, Hideyo, 151
Nomadic peoples, 22-23, 25-26, 48-49
    arrival in Americas, 97
Nome, diphtheria epidemic, 146
Nongonococcal urethritis (NGU), 206
North Africa, bubonic plague in, 89, 90
North America, 91, 104, 202. *See also* United States
North Dakota, hantavirus pulmonary syndrome in, 172
Norwalk virus, 219
Nutrition, and health, 140-41
Nutritional diseases, 35-36, 98, 99, 107

Occupational diseases, 34-35, 51-52
Oceania, syphilis in, 123
Old Lyme, Connecticut, 176
"Old Man of La Chapelle-aux-Saints," 26
Old Testament, 55-56, 69, 83
Oman, polio in, 154
Omsk hemorrhagic fever, 168
On Contagion (Fracastoro), 115
Oncogenes, 201
Oncoviruses, 199-200, 201
O'nyong-nyong fever, 5, 6, 168
Oral herpes, 203
Organ transplants, 224
Oropouche, 5, 6, 162
Orosius, plague of, 70
Overpopulation, 68, 76. *See also* Population growth
Oviedo y Valdes, Gonzalo Fernández de, 125
Oxen, 37

Pacific islands, 31, 110
"Pale Horse, Pale Rider" (Porter), 145
Paleolithic hunters, 34
Paloma, Peru, 33, 35
Panama, Oropouche epidemic, 162
Pandemics, 18, 65-66, 68-69, 110, 226
    AIDS, 185-94
    bubonic plague, 75-78, 86-91

cholera, 132-38, 215
    influenza, 38, 143
    sexually transmitted diseases, 205-7
    slow viruses, 201
    twentieth century, 143
Panspermia theory, 15-16, 83
Papua New Guinea, 197-98
Paragonimus, 33
Parasitism, 10, 13, 16-18
    agriculture and, 42
    domestic animals and, 36, 37
    in tropical regions, 22, 25
    See also Intestinal parasites
Paris, 133, 135, 136
Parkinson's disease, 200
Parmenter, Robert, 172
Parvoviruses, feline and canine, 223
Pasteur, Louis, 138-39
Pathogens, mutation of, 141
Pavlovsky, Evgeny, 39
PCP. *See Pneumocystis carinii*
Peloponnesian War, 59-60
Pelvic inflammatory disease, 206
Pepys, Samuel, 90
Persia, 124, 132
Pertussis, 6
Peru, 97, 102, 103, 104, 118, 216
    Paloma, 33, 35
Pesticides, 155
    resistance to, 227
Peter II, Tsar of Russia, 109
Petrarch, 89
Philadelphia, 107, 117, 152, 170
    Legionnaires' disease in, 181-83
    1918 influenza pandemic, 144, 145
Philippine Islands, 43-46, 159-60
Phytophthora infestans (potato blight), 118
Picardy sweats, 81
Pigs, 36, 38, 39
Pinta, 100, 126, 127, 128
Pituitary growth hormone, 200
Pizarro, Francisco, 102
Plant diseases, 118
Plasmids, 212
Plasmodium falciparum, 40
Plasmodium ovale, 40
Plowing, and disease, 41
Plutarch, 83
Pneumocystis carinii, 185, 224
Pneumonia, 100, 144, 217, 224

Pneumonic plague, 2, 6, 75, 77, 88
Pogroms, 88
Poland, 89, 123, 124, 132
Polar bear meat, raw, 23
Poliomyelitis, 53, 149-54
Political refugees, 218
Pollution, 4, 41-42, 52, 53, 129, 131
Polynesia, European diseases in, 110
Pontiac fever, 6, 7, 183
Population control, 67
Population decline, 65-66, 86
Population density, and epidemics, 49
Population growth, 48-50, 69, 80, 94,
    113, 119, 130, 221, 224-25
  agriculture and, 41
  diseases and, 42-43, 61-62
    epidemics, 65, 86, 194
    HIV, 190
    sexually transmitted, 207
  and environmental change, 156
  health conditions and, 140-41
  twentieth century, 142, 155
Porogia virus, 169
Porotic hyperostosis, 35
Porter, Katherine Anne, 145
Portugal, 95, 124
Potato famine, Ireland, 118-21
Pott's disease, 84
Poultry, 36, 37, 39
Poverty, 207, 209-10
Pox infections, in animals, 38
Predation, self-limitation, 16
Prehistoric animals, 14
Prehistoric diseases, 18-20
Prehistoric peoples, 32-33
Premature infants, 224
Primates, diseases of, 19-20
Prion protein molecule, 200
Procopius, 74
Progress, scientific, faith in, 3
Prospect Hill hantavirus, 170
Prosperous people, sexual
    behavior, 207
Prostitution, 185, 191, 192
Protein-poor diets, 35-36
Prowazek, Stanislaus von, 114
Prusiner, Stanley, 200
Prussia, cholera in, 135
Pubic lice, 113-14
Public health, nineteenth century,
    137-38

Public health systems, 134, 217-19
Pudoc, Philippines, mystery disease,
    43-46
Puerto Rico, dengue in, 159, 160
Puumala virus, 169, 172

Q fever, 42
Quarantines, 88, 134-36, 223
  of laboratory animals, 167
Quartan malaria, 107

Rabbits, 24, 36
Rabies, 9-10, 36, 138, 139
Rain forests, 62, 190
Rash diseases, 57-59
Rat(s), 39, 68, 170
  and bubonic plague, 24, 76-77, 87
Rat fleas, and typhus, 113-14
Recovery, environmental, 179-81
Reed, Walter, 139, 151-52, 158-59
Reforestation, 179-80, 220
Reformation, plague and, 90
Relapsing fever, 24, 52, 120
Renaissance, 86, 90
Research, in polio, 152-53
Reservoirs of disease, 8, 156, 158
Respiratory equipment, 184
Respiratory infections, 122
Reston virus, 167
Retroviruses, 187-88, 190,
    199-200
  and cancer, 201-2
Revelation (Bible), 48
Reverse transcriptase, 187
Revolution, 30
Rhazes, 58, 71
Rheumatic fever, 211
Rheumatoid arthritis, 5
Richard III, King of England, 81
Rickets, in early man, 35
Ricketts, Howard, 114
Rickettsiae, 15, 42, 54, 113-14, 122
Rift Valley fever (RVF), 6,
    163-64, 220
Ritual cannibalism, 198
River blindness, 37
RNA, 14, 187
Robinson, Edward G., 139
Rocio virus, 157
Rockefeller Foundation Virus
    Program, 162

Rocky Mountain spotted fever (RMSF), 4, 42, 177, 220
Rodents, 42, 156, 161-63
  and bubonic plague, 24, 76-77, 86-87
  See also Mice; Rat(s)
Roman empire, 66, 69, 72-75
Rome, 89
  ancient, 58, 62, 69-72, 85
Roosevelt, Franklin, 150
Rosebury, Theodor, 125
Ross River fever, 168
Rotavirus, 219
Rous sarcoma virus, 201
Rubella, 58
Rush, Benjamin, 107
Russia, 89, 116, 117, 155, 169
  cholera in, 132, 135
  influenza pandemics, 143, 144
  syphilis in, 123, 124
Russian spring-summer encephalitis, 158
RVF (Rift Valley fever), 163-64
Rwanda, 217

Sabin oral vaccine, 153
Safe sex practices, 202-3
Sailors, and syphilis, 123
St. Elizabeth Hospital (Washington, D.C.), 183, 184
St. Louis encephalitis (SLE), 7, 157
Salk vaccine, 153
Salmonella, 24, 37, 53, 207, 220
Salvarsan, 139
Samoa, diseases in, 59, 144
San Diego, legionellosis in, 184
Sanitary reform movement, 137
Sanitation, 45, 137, 155
  and polio virus, 151-52, 153
Santo Domingo, 103-4, 133
Satellite sensing imagery, 164
Scandinavia, 35, 50, 84
Scarlet fever, 48, 57, 58, 103, 108, 141, 211
Schistosomiasis, 41-42, 62, 74, 229
Schizophrenia, 201
Science, nineteenth century, 134
Scopes trial, 30
Scotland, syphilis in, 123
Scrapie, 196-99
Scrofula, 84

Scrub typhus, 24, 42, 117
Scrub vegetation, and disease, 42
Scurvy, 23, 119
Sedentism, 34
Selye, Hans, 68
Seoul hantavirus, 6, 10, 169, 170
Septicemic plague, 88
Seven Years War, 116
Sexual revolution, 205
Sexually transmitted diseases (STDs), 9-10, 53, 122-23, 205-7, 218
  cervical cancer as, 205
  giardiasis, 219
  hemorrhagic fevers, 158
  hepatitis B virus, 202
  HIV infections and, 191-92
  syphilis, 123-28
Shanidar, Iraq, 27
Sheep, 37, 39, 196-97
Shigellosis, 42, 53, 138, 207, 220
Shingles, 203
Ships, and spread of disease, 222
Shope, Robert, 228
SHV (simian hemorrhagic virus), 167
Sibbens, 127. *See also* Bejel
Siberia, 109-10, 123, 124, 144
Sickle-cell trait, 40
Sieges, typhus and, 114
Siena, bubonic plague in, 89
Sierra Leone, Lassa fever in, 164
Sigurdsson, Björn, 197
Simian AIDS (SAIDS), 187
Simian hemorrhagic virus (SHV), 167
Simian immunodeficiency virus (SIV), 189
Sindbis, 158
Single-crop farming, 86, 118
Sin Nombre, 10, 173
Sixteenth century, 94, 143
Skin diseases, 52-53
Slavery, 90, 95, 106-7
SLE (St. Louis encephalitis), 157
Sleeping sickness, 21, 24, 38, 41, 43, 54, 229
Slow viruses, 2, 197-200
Smallpox, 11, 38, 48, 57, 58, 71-72, 73, 101-5, 108, 109, 110, 226
  eradication of virus, 154-55
  inoculation, 141-42
Smolensk, bubonic plague in, 89
Snow, John, 138

Social change, 130, 190
Social chaos, disease and, 43, 217
Social class, 51
    cholera and, 133, 134–35
Social customs, and spread of
    infections, 54
Social development, 42, 98
Soldiers, and syphilis, 123
Solecki, Ralph, 27–28
Somalia, 217
South Africa, influenza in, 144
South America, 159, 160, 161–62,
    180, 215–16, 220
Southeast Asia, 42, 159–60, 202
    cholera in, 132, 217
Southern United States, malaria in, 109
"Souvenir species," 37
Soviet Union, former republics, 217
Spain, 95, 115
Spanish conquest, New World, 102, 103
Spanish flu, 143, 144. See also
    Influenza
Spirochetes, 126
Spongy encephalopathies, 198–99, 200
Sprent, J. F. A., 37
Springfield, Massachusetts, 219
Spumiviruses, 200
Staphylococcus infections, 9, 19, 24,
    210–11
Starvation, 140
Staten Island, New York, polio
    epidemic, 152
Steere, Allen, 176
Stewart, William H., 3
Stimulation, human, urban growth
    and, 63
Stockholm, cholera in, 133, 136
Stomach cancer, 205
Stone Age societies, study of, 32–33
Streptococci, 19
    type A, 9, 211, 226
Streptomycin, 141
Stress, physical effects, 68
Stress fractures, Neolithic, 35
Strokes, hantaviruses and, 170
Suburbia, and disease, 179–80, 220
Sudan, diseases in, 5, 166
Sugar cane, 95
Sulfonamides, 139, 141
Surgical instruments, and Creutzfeldt-
    Jakob disease, 200

Sweating sickness, 81
Sweden, 123, 150
Swine flu vaccine, 8
Switzerland, syphilis in, 123
Symbiosis, 18
Syphilis, 4, 5, 6, 48, 53, 57, 109,
    110, 111, 121–28, 193, 205
    and HIV, 207
    in precontact Amerindians, 100
    treatment for, 139
Syphilis sive morbus gallicus
    (Fracastoro), 124
Syracuse, ancient, plague of, 69
Systemic lupus, 5

Tagudin, Philippines, intestinal
    capillariasis in, 44–45
Tapeworms, 25, 37, 38
Technology, 50, 94–95, 110, 134,
    223
    and AIDS, 190
    Amerindians and, 97
    and cholera, 130
    and Legionnaires' disease, 184
    and oncoviruses, 201
Tell Abu Hureyra, Syria, 33–34
Tenochtitlán, Mexico, 101
Tertian malaria. See Malignant tertian
    malaria
Tetanus, 19, 24, 139
Texas, 172, 180
Thailand, 44–45, 170
Thalassemia, 40
Thames River, pollution of, 136
Third World, 62, 207
Thirty Years War, 115
Thucydides, 59, 60
Tick-borne diseases, 24, 36, 42, 54,
    168, 177, 180–81
    Lyme disease, 8, 176–80
Tigris-Euphrates valley, 30
Tissue transplants, 223–24
Tobacco mosaic virus, 118
Toxic-shock-like syndrome (TSLS), 6,
    9, 211, 226
Toxic shock syndrome (TSS), 6, 9,
    211, 226
Toxoplasmosis, 24, 52, 185, 224
Trachoma, 206
Trade, 61, 70–71, 85–86, 87, 222–23
    in laboratory animals, 164, 191

Transmission of diseases, 11, 16–
17
arboviruses, 156
in cities, 52
HIV, 190
polio, 153
tuberculosis, 84, 208
and virulence, 18
Travel, 7, 60, 61, 192, 222–23
Trench fever, 117
Treponemes, 126–28
Triatoma infestans, 98
Trichinosis, 23–24, 37, 190
Trinidad, dengue epidemic, 159
Tropic regions, parasitism in, 22
Tryde, E. A., 23
Trypanosoma brucei, 38
Tsara'at, 83
Tsetse flies, 19, 21
Tuberculosis, 5, 6, 11, 20, 42, 53,
100, 109, 110, 122, 139, 141,
208, 218, 229
and AIDS, 185, 224
in cave bears, 19
drug-resistant, 2
and leprosy, 82, 84–85
Tularemia, 24, 190
Tumors, viruses and, 201
Turkey, syphilis in, 124
Twentieth century, 3, 142, 150
influenza pandemics, 144–47
Typhoid, 53, 138, 217, 218
in New World, 103, 107, 108
"Typhoid Mary," 152
Typhus, 3, 48, 52, 54, 110, 111, 113–
17, 128, 137, 139, 217
famine and, 118–21
in New World, 103, 108
twentieth century pandemic, 143

Uganda, 5
Ulcers, 205
Undercooked foods, 220
Undulant fever, 38
Unicorn (Danish ship), 23–24
United States, 24, 37, 42, 138, 157
202, 218, 219–21
AIDS in, 185
cholera in, 133, 215–17
and dengue, 159, 160
and eradication of smallpox, 155

influenza epidemic, 144, 145
Irish immigrants, 120
malaria in, 41, 108–9
polio epidemics, 150–53, 154
sexually transmitted diseases, 191
chlamydia, 206
syphilis, 205–6
tuberculosis in, 208–9, 210
yellow fever in, 158–59
United States Army Medical Research
Institute of Infectious Disease
(USAMRIID), 167, 171
Universities, U.S., 108
Urban environments, 50–51, 55, 84
Urban growth, 62–63
Urban societies, 33–35, 40–41
Urbanization, 94, 192, 220–21, 225
twentieth century, 142
Ussher, James, 30

Vaccination, 141–42, 218
Vaccines, 139, 155
BCG, for tuberculosis, 208
for hepatitis B virus, 202–3
Valley fever, 217
Vancouver, George, 105
Vectors of disease, 53–54, 156
Venezuela, syphilis in, 127
Venezuelan equine encephalitis, 157
Venezuelan hemorrhagic fever, 6, 163
Venice, bubonic plague in, 88
Vertical zones, environmental, 20
Vibrios, 131–39, 215–17
Villages, 43, 49, 55
Viral infections, 49
encephalitis, 6, 7, 49, 157
hepatitis, 6, 10, 53
Virchow, Rudolf, 19
Virginia, colonial, 107
Virologists, 167–68
Virulence of disease, 16, 18, 23
changes in, 212
Viruses, 5, 14–16, 17, 151, 187
and cancer, 201–2
Visna virus, 197, 200
Vitamin deficiency diseases, 35–36

Wallin, Ivan, 15
War, 98
disease and, 43, 60–61, 71, 217
typhus, 114–15, 116–17

Washington, George, 42
Washington, D.C., 167, 170
Water, 129, 136
  contaminated, 37, 53, 138
Water systems, 219
  Legionnaires' disease in, 184
Waterborne infections, 53, 218–19
  cholera, 129–38
  sanitation movement and, 137–38
Welfare, nineteenth century, 121
West Africa, 59, 158
West Indies, smallpox in, 101
West Nile fever, 157
Western equine encephalitis
  (WEE), 157
White, David, 13–14
WHO (World Health Organization),
  154
Whooping cough, 5, 53, 108,
  110, 218
Wild game, diseases from, 24–25
Wilson bands, 33
Winthrop, John, 105
Wolfe, Thomas, 47
World Health Organization (WHO),
  154
World population, Christian era, 69
World War I, 3, 142–43, 144, 145
  typhus epidemic, 117

World War II, 3, 42, 155, 160, 161
  and typhus, 117
Worms, parasitic, 21
Wound infections, 19, 100, 211

Xenopsylla cheopis, 77

Yaws, 20, 126, 127–28
Yellow fever, 19, 24, 38, 40, 52, 54,
  133, 139, 156, 158, 217
  Asian tiger mosquitoes and, 160
  environmental change and, 20
  in New World, 106–7, 108
Yersin, Alexander, 76
Yersinia pestis, 76–78, 86–87
Young, Delbert, 23–24
Young people, 207–8, 225
Yucatán, yellow fever epidemic, 106
Yugoslavia, babesiosis in, 180

Zaire, 5, 166, 168–69, 218
Zebra mussels, 222
Zero Population Growth, 67
Zimmerman, Michael, 22
Zinsser, Hans, 117
Zoonoses, 23–24, 32, 36–39, 51, 53,
  126, 168, 172, 219
  HIV as, 190
  new, 56–57
Zymotics. See Crowd diseases